# Collineations and Conic Sections

Christopher Baltus

# Collineations and Conic Sections

## An Introduction to Projective Geometry in its History

 Springer

Christopher Baltus
Department of Mathematics
State University of New York at Oswego
Oswego, NY, USA

ISBN 978-3-030-46289-5        ISBN 978-3-030-46287-1    (eBook)
https://doi.org/10.1007/978-3-030-46287-1

Mathematics Subject Classification: 01A05, 51-03, 51A05, 51N15, 01A45, 01A55, 97G50

This Springer imprint is published by the registered company Springer Nature Switzerland AG
The registered company address is: Gewerbestrasse 11, 6330 Cham, Switzerland

*My thanks to SUNY Oswego and its Department of Mathematics for their support; to my wife, Banna Rubinow, for her patience, encouragement, and editing; and to the publisher, Springer, for supporting a book outside the usual categories, combining mathematics and its history.*

*Christopher Baltus*

# Preface

Searching for an entry into the projective geometry section of a college geometry course I was teaching many years ago, I found central collineations in Dan Pedoe's *Introduction to Projective Geometry*. In a central collineation, the plane is mapped to itself so lines go to lines and there is a line of points unchanged by the collineation. Here was a simple idea which led to lovely results in proofs that were actually fun and easily motivated by diagrams.

The next discovery, for me, was Philippe de la Hire's projective introduction to the conic sections, from 1673, to which he added some 20 pages that he called *Plani-conique*. *Plani-conique* is a set of simple rules to carry out a plane-to-plane projection when one plane is imposed on the other. These are exactly the rules of the central collineation!

Moving ahead 140 years, there is the story of Jean-Victor Poncelet. Captured while Napoleon's army made its wintry retreat from Moscow in late 1812, as a prisoner he rebuilt the geometry he recalled from student days at the *École Polytechnique*, without the benefit of books. He made great use of what we now call dilations, and from the notebooks he wrote as a prisoner and his 1822 treatise on projective geometry, you could see him discovering, again, central collineations. (I believe the place of the central collineation in the work of Poncelet is not widely recognized.)

An idea this central, with interesting problems, accompanied by straightedge constructions, makes a good story that deserves wide awareness. Central collineations provide a rich approach to projective geometry, an approach that is little developed beyond Pedoe's book and Luigi Cremona's *Elements of Projective Geometry* of 1893. I have combined an introduction to central collineations with an introduction to projective geometry, set in its history.

Except for a couple of sections that presume background in calculus and matrix algebra, and some modular arithmetic, the prerequisite knowledge is high school geometry. This is a proof-based treatment, so a measure of enthusiasm for logical thinking, accompanied by diagrams, is the other prerequisite. Some of the proofs are challenging, but secondary mathematics is sufficient background. There are also sections to give the reader, who I think of as junior-year mathematics major, the background needed by someone who has not taken a college geometry course. In fact, this book could serve as a college geometry course that emphasizes projective geometry.

The history is selective, concentrating on a few characters and just one or two works by each, often works early in their careers, so a reader will be able to learn the mathematics along with developing a good sense of its history. I have brought in central collineations wherever there is good reason to, but, recognizing that no one strand does justice to the subject of projective geometry, this book aims to provide the reader a general history through the middle of the nineteenth century.

Optional material, which can be skipped without limiting a reader's ability to work through the rest of the book, is set aside in excursion sections. The exercises can be skipped by readers whose interest is in the history.

Behind this work is the conviction that combining history and a mathematics exposition is a good way to learn both. Where history and historiography might bog down the mathematics, I have put information into chapter notes, where I have also included exercises. Several claims not crucial to the story, with their proofs, have also been placed with the notes.

The synthetic development of projective geometry is no longer an active area of research. ("Synthetic" means in the tradition of Euclid's *Elements*, without coordinate geometry.) For projective geometry, in the last century, foundations issues and applications in algebraic geometry have dominated. On the other hand, as a window into the place of mathematics in cultural history, synthetic projective geometry is ideal. Further, it has accessible proofs and beautiful constructions and in that way is an excellent entry to theorem-proof mathematics, both as a way to improve one's skills and as a way to see how theorem-proof mathematics grew over the centuries to elaborate the amazing properties that follow from simple plane-to-plane projection.

I end this preface with a quote from J. L. Coolidge's 1933 address with the dispiriting title "The Rise and Fall of Projective Geometry."

> ... the field of synthetic projective geometry is pretty much worked out. ... On the other hand ... [synthetic methods not only have] a permanent beauty which no one who has ever studied them can forget, but they afford an invaluable insight into the inner significance of geometric science and an invaluable training for any geometer. If Plato wrote over the gate of the Academy, "Let none ignorant of geometry presume to enter here," surely we may write today in the same spirit, "Let none ignorant of the fundamentals of synthetic projective geometry presume to the title of geometer." [*American Mathematical Monthly*, 1934, p 227]

Oswego, NY, USA                                                       Christopher Baltus

# Contents

# Chapter 1
# Introduction: The Projective Plane and Central Collineations

This book grew from the idea that much of projective geometry is the elaboration of a simple concept, the central collineation. A central collineation is a construction, carried out with just a straightedge and a device to construct parallel lines, following the simplest rules, that transforms one diagram into another. We call it a *collineation* because a line is always transformed to a line. We'll soon explain the meaning of *central*. And how is a circle transformed? That is the marvelous part, for a circle becomes a parabola or ellipse or hyperbola—a conic section. And all conic sections can be formed this way.

## 1.1 The Projective Plane

To take full advantage of the central collineation, one needs to work in its most natural arena, the projective plane. A projective plane is an extension of an *affine plane*. The affine plane that we will generally deal with is the real affine plane, $\mathbf{R}^2$, the plane of secondary mathematics which, since the time of Descartes, we have thought of as the set of ordered pairs of real numbers $(x, y)$. There, lines are the solution sets of equations $ax + by + c = 0$ when $a, b, c$ are real numbers and $a$ and $b$ are not both zero.

We extend an affine plane to a projective plane by assigning an additional point, which we say lies at infinity, to each line of the affine plane, but in such a way that parallel lines share the same additional point at infinity. And then the entire collection of points at infinity is, itself, designated as a line.

Renaissance artists learned to consider parallel lines in a plane as meeting on the line at infinity of that plane, and in representing that plane in their paintings, they had a horizon line at which the parallel lines meet.

Figure 1.1 is from an instruction book for artists, published in 1652. The illustration is by Jan Vredeman de Vries, who has kindly labeled the horizon line

© Springer Nature Switzerland AG 2020
C. Baltus, *Collineations and Conic Sections*,
https://doi.org/10.1007/978-3-030-46287-1_1

**Fig. 1.1** Vredeman, 1605

*Orizon*, the line on which any pair of lines parallel to each other and parallel to the plane of the floor meet [6].

It may not be a coincidence that the first work of projective geometry, of 1639, by Girard Desargues, was written just decades after the first works to treat perspective drawing as a science, and came just 3 years after Desargues himself had written a pamphlet on perspective drawing.

## 1.2   Homogeneous Coordinates and the Real Projective Plane

To see precisely how points can be added to the plane of real ordered pairs, in a way that any two lines meet in one point, we introduce *homogeneous coordinates*, developed in the nineteenth century.

The basic elements of a geometry in a plane are points and lines. In homogeneous coordinates, both points and lines are represented by triples of real numbers. As we will come to see, *duality* is expected in projective geometry. The *dual* of "any two points lie on exactly one line" is the statement "any two lines lie on exactly one point," in which the roles of line and points are reversed. So if lines are represented by triples, that should also be the case with points.

Let us start with lines.

In the real affine plane, the plane of high school geometry, lines are the sets of solutions $(x, y)$ of equations $ax + by + c = 0$, where $a$ and $b$ are not both 0. So a line can be represented as a triple $(a, b, c)$ where $a$ and $b$ are not both 0. This collection of lines has the *homogeneity* property: triples $(a, b, c)$ and $(ka, kb, kc)$ represent the same line for any non-zero $k$, since equations $ax + by + c = 0$ and $kax + kby + kc = 0$ have the same solution set.

Now points. $(x, y)$ satisfies $ax + by + c = 0$ exactly when triple $(x, y, 1)$ satisfies $ax + by + cz = 0$. So a point $(x, y)$ of the affine plane is to be represented as point $(x, y, 1)$ in homogeneous coordinates. And because the triple $(x, y, z)$ satisfies $ax + by + cz = 0$ when, for $k \neq 0$, triple $(kx, ky, kz)$ satisfies $ax + by + cz = 0$, then points also satisfy the homogeneity property: for $k \neq 0$, $(x, y, z)$ and $(kx, ky, kz)$ represent the same point. Points that can be represented in form $(x, y, 1)$ are *finite points*. When $z \neq 0$, then any point $(x, y, z)$ can be expressed, by the homogeneity property, in form $(x, y, 1)$, corresponding uniquely to point $(x, y)$ of the affine plane.

What about the *points at infinity* we wish to add to form a projective plane? We let $z$ be 0. What happens when $z = 0$? Point $(x, y, 0)$ cannot be expressed in form $(x, y, 1)$, so it is not a finite point. Point $(x, y, 0)$ lies on line $(a, b, c)$ exactly when $ax + by = 0$. (We do not allow a point for triple $(0, 0, 0)$ since it would lie on all lines.) Let us fix the values of $a$ and $b$, not both 0. Then the set of lines $(a, b, c)$, for all values of $c$, is the set of lines parallel to line $(a, b, 0)$, for two such lines meet only at point $(b, -a, 0)$, not a finite point.

The set of all points of form $(x, y, 0)$, where $x$ and $y$ are not both 0, forms the *line at infinity*, represented by $(0, 0, 1)$ in homogeneous coordinates.

**Theorem 1.1** *Any two lines, as defined in homogeneous coordinates, meet in exactly one point.*

**Proof** We consider two lines $(a_1, b_1, c_1)$ and $(a_2, b_2, c_2)$, with $a_1$ and $b_1$ not both 0, $a_2$ and $b_2$ not both 0. We may suppose that $a_1 \neq 0$. Then the system of two lines

$$a_1 x + b_1 y + c_1 z = 0$$

$$a_2 x + b_2 y + c_2 z = 0$$

can be represented

$$ka_1 x + kb_1 y + kc_1 z = 0$$

$$a_2 x + b_2 y + c_2 z = 0$$

for $k = \dfrac{a_2}{a_1}$, and subtracting replaces the system by one of form

$$b_3 y + c_3 z = 0$$

$$a_2 x + b_2 y + c_2 z = 0.$$

(Take $b_3 \neq 0$ since $b_3 = 0$ when the lines are parallel, in which case they meet at a single point on the line at infinity.)

When $c_3 \neq 0$, then $z$ cannot be 0, since $(0, 0, 0)$ is not a point in homogeneous coordinates. In this case, we take $z = 1$, and then $y$ and $x$ are uniquely determined by the two equations. And when $c_3 = 0$, then $y = 0$ and $z$ cannot be 0; we can take $z = 1$ and $x$ is determined by the equation of the first line. We thus have the unique point lying on the two given lines.

The reader can check that line $(0, 0, 1)$ meets any other line in exactly one point.

In summary, we have                                                                                                        □

**Definitions I** The *real projective plane*, which we denote $\mathbf{P}^2$, is the set of points $\{(x, y, z) :\ x,\ y,\ z\ are\ real\ and\ not\ all\ 0\}$, where the lines are the triples $(a, b, c)$ with $a, b, c$ real and not all 0. Point $(x, y, z)$ lies on line $(a, b, c)$ exactly when $ax + by + cz = 0$.

Except where noted, "the projective plane" refers to this real projective plane $\mathbf{P}^2$. At this point, the reader is alerted to the existence of other projective planes, including planes of finitely many points. This more abstract view of projective geometry emerged in the nineteenth century.

Then, one might ask what makes a projective plane "projective"? Here we list the most essential properties of both affine and projective geometry, and then, in our first Excursion, we turn to the smallest examples of finite projective and affine planes. In the case of the real affine plane and of the real projective plane, the properties can be proved by coordinate geometry; in other geometries they may be assumed as axioms.

### Properties of an Affine Plane

*Property A1*  Any two points lie on exactly one line.

*Property A2*  (Parallel Postulate) Given a line $m$ and a point $X$ not on $m$, there is exactly one line on $X$ which does not meet $m$.

### Properties of a Projective Plane

*Property P1*  Any two points lie on exactly one line.

*Property P2*  Any two lines lie on exactly one point.

*Property A3*  In an affine plane, there exist at least three points not all collinear, and every line has at least two points.

*Property P3*  In a projective plane, there exist at least four points of which no three points are collinear, and every line has at least three points.

*Property* A3 and *Property* P3 guarantee we have enough points and lines for a meaningful geometry. It can be shown that Properties P1 and P2 and the first part of A3 imply the second part of A3 (See Exercise 6)

## 1.3   Central Collineation: Definition and Elementary Properties

We have set out the details, the points and lines, of the arena in which we will work, a projective plane. But what happens in that arena?

We transform it. We shall use the term *transformation* for a function defined on a projective plane that is *one-to-one* and *onto* the same projective plane. Typically, a function rule gives the image point for each input point. However, this is projective geometry, where duality reigns, so any function should also, alternatively, give the image line for each input line. This means we only wish to consider transformations which map lines to lines. Such functions are called *collineations*.

And we will emphasize a type of collineation known as a *central collineation*. We develop a central collineation by starting from a plane-to-plane mapping, from plane $\pi$ to plane $\pi'$, in three dimensional space, as in Fig. 1.2.

The two planes, $\pi$ and $\pi'$, meet in a line $l$, called the *axis*, and there is a point $O$ which lies in neither plane. We could think of $O$ as an artist's eye, traditionally labeled $O$, perhaps after the French *oeil*, for *eye*. One plane could be the plane of the artist's canvas and the other an endless tile floor, although, in our example, we would be projecting the canvas onto the floor. Then point $X'$ of plane $\pi'$ is the image of point $X$ of plane $\pi$ when $O$, $X$, and $X'$ are collinear. We make the following observations.

**Fig. 1.2**  Projection in space of plane $\pi$ to plane $\pi'$

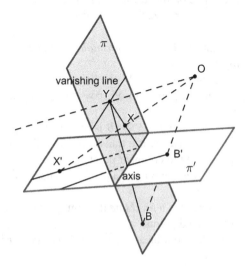

1. The points of line $l$ are mapped to themselves. So $l$ is a *line of fixed points* or *axis*.
2. This map is one-to-one: two points of $\pi$ cannot have the same image since two lines on $O$ cannot meet again on plane $\pi'$.
3. Collinear points are mapped to collinear points. For if we take a line $m$ in $\pi$, its image, line $m'$, must lie in the plane on $O$ and line $m$, and this plane meets plane $\pi'$ in a line, $m'$. In this sense, our mapping is a *collineation*.
4. If $Y$ lies in $\pi$ so that line $OY$ is parallel to plane $\pi'$, then there is no finite $Y'$ in $\pi'$. If $q$ denotes the line in $\pi$ on such a point $Y$ that is parallel to the axis, $l$, then no point of $q$ has a finite image in $\pi'$. $q$ is called the *vanishing line*.
5. Two lines, $m$ and $n$, of $\pi$ which meet at a point $Y$ of $q$ are mapped to parallel lines in the affine plane $\pi'$ since there is no finite point at which those image lines could meet. We declare that $Y'$ is to be the point at infinity of lines $m'$ and $n'$ in plane $\pi'$. All such points $X'$, the images of points $X$ on $q$, form the *line at infinity of plane $\pi'$*.
6. There also is the line $r'$ of plane $\pi'$, not marked in the diagram, where the plane on $O$ that is parallel to plane $\pi$ meets plane $\pi'$. $r'$ has no finite pre-image in $\pi$. We can say that the line at infinity in $\pi$ is the preimage of $r'$

Thus we have a one-to-one mapping of the projective plane $\pi$ onto the projective plane $\pi'$, where lines are mapped to lines.

Next, we rotate plane $\pi'$ about line $l$—either way—until it lies on plane $\pi$. Now each point $X$ and its image $X'$ lie in one plane. We have a one-to-one and onto collineation of plane $\pi$ to itself, which includes a line, $l$, of fixed points.

After one plane is rotated about the line on which the two planes meet, so that $\pi = \pi'$, we have a central collineation of a plane onto itself. The feature that makes this collineation "central" is that it has an axis, a line of fixed points. As we shall see, the presence of an axis is equivalent to the presence of a *center*, a point where every line on that center is mapped to itself.

We summarize with definitions.

**Definitions II**  A *transformation* is a one-to-one mapping of a projective plane onto itself.

A function $f : A \rightarrow B$ is *one-to-one* if different points are always mapped to different points. In other words, for all pairs of different points $x_1$ and $x_2$ in $A$, $f(x_1) \neq f(x_2)$.

A function $f : A \rightarrow B$ is *onto* $B$ if every $y$ in $B$ is $f(x)$ for some $x$ in $A$.

A *collineation* is a transformation of the projective plane where collinear points are mapped to collinear points.

Under a mapping of a plane onto itself, a point $X$ is a *fixed point* if its image, $X'$, is $X$ itself.

A line $m$ is a *fixed line* if its image, $m'$, is $m$ itself. The points of a fixed line need not be fixed points.

Under a collineation, a point $V$ is a *center* if every line on $V$ is a fixed line.

Under a collineation, a line of fixed points is called an *axis*.

A collineation which has a center is called a *central collineation*.

Line $m$ is the *vanishing line* if it is mapped to the line at infinity.

When a collineation has a line of fixed points, $l$, then it has the dual of $l$, a *center*. This is Lemma 1.4 in this upcoming sequence of lemmas.

**Lemma 1.1**

(a) *If a collineation has two lines of fixed points, then it is the identity map, and, the dual,*

(b) *if a collineation has two centers, then it is the identity map.*

*Proof in the Case of Two Lines of Fixed Points*   See Fig. 1.3. Let $CD$ and $AH$ be two lines of fixed points for a collineation $\phi$. Let $P$ be a point on neither $CD$ nor $AH$. Take two lines on $P$, $EH$ and $FG$, with $E$ and $F$ on $CD$, and $G$ and $H$ on $AH$. These lines are fixed by $\phi$ since each has two fixed points. So $P'$ lies on both $EH$ and $FG$, so $P' = P$. So all points of the projective plane are fixed by $\phi$. The proof of $b.$ is left as an exercise.

**Lemma 1.2**  *If a collineation $\phi$ has a line, $l$, of fixed points, then any fixed point, $P$, not on $l$, must be a center.*

**Proof**  Suppose $m$ is a line on a fixed point, $P$. $m$ must meet $l$ in a fixed point, $L$. $m'$ must lie on $P$ and on $L$, so $m = m'$.                                 □

**Lemma 1.3**  *If a collineation $\phi$ has a line, $l$, of fixed points, and $A$ is not a fixed point, then line $AA'$ is a fixed line.*

**Proof**  Line $AA'$ meets line $l$ in a point $K$, which is itself fixed. Since points $A$ and $K$ are mapped to points on line $AA'$, then line $AA'$ is mapped to itself.       □

**Lemma 1.4**  *If a collineation $\phi$ has a line, $l$, of fixed points, then it has a center.*

**Fig. 1.3**  Proof of
Lemma 1.1, case of two lines
of fixed points

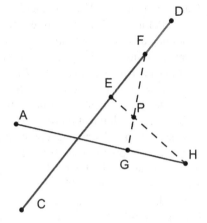

**Fig. 1.4** Proof of Lemma 1.4

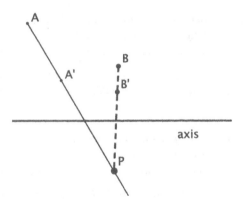

*Proof of Lemma 1.4* See Fig. 1.4. Suppose there is no center on $l$. If $\phi$ had a fixed point $P$ not on $l$, then by Lemma 1.2, $P$ would be a center. So now suppose there is no center on $l$ and no fixed point off $l$. Take a point $A$ not on $l$ and another point, $B$, on neither $l$ nor line $AA'$. ($B'$ is on neither $l$ nor on $AA'$. Why ?) Let $AA'$ and $BB'$ meet in point $P$. Then $P$ is a fixed point. So $P$ is a center. [Since there is no center on $l$, there is such a line $BB'$ that meets line $AA'$ off of $l$.]                    □

## 1.4 Excursion: Finite Affine and Projective Planes of Minimum Size

To create an affine plane of minimum size, we use a field, denoted $F_2$, with just two elements, 0 and 1, where addition and multiplication are carried out *mod* 2. The plane is simply the set of the four ordered pairs $\{(x, y) : x, y, \in F_2\}$; the lines are the solution sets of equations $ax + by + c = 0$ when $a, b, c$ are in $F_2$, and $a$ and $b$ are not both zero. You find there are six lines, of two points each, in three pairs of parallel lines. We form a projective plane, first developed by Gino Fano in 1892, by adding a "point at infinity" to each pair of parallel lines, and declaring the three points at infinity to, themselves, form a line.

You can check that Properties *P1., P2.,* and *P3.* are satisfied.

Exercise 5, at the end of this chapter, asks the student to form an affine plane of all the ordered pairs of numbers in field $F_3$, where $F_3$ is the set $\{0, 1, 2\}$ under addition and multiplication (mod 3). And then that affine plane is to be extended to a projective plane by adding a point at infinity for each collection of parallel lines.

In Fig. 1.5, points $A = (0, 0)$, $F = (1, 0)$, $C = (1, 1)$, and $G = (0, 1)$ form an affine geometry, with lines $AF : y = 0$, $AC : x + y = 0$, $AG : x = 0$, $GC : y = 1$, $FG : x + y = 1$, and $FC : x = 1$. We can check that Properties *A1.* and *A2.* are satisfied. The pairs of parallel lines are simply those that share no common point. In homogeneous coordinates, we have $A = (0, 0, 1)$, $F = (1, 0, 1)$, $C = (1, 1, 1)$, and

**Fig. 1.5** Fano's Seven Point Projective Geometry. The points on the circle constitute a line

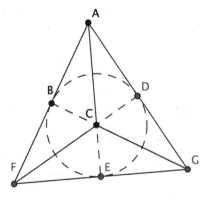

$G = (0, 1, 1)$. The parallel pairs meet in points of the "line at infinity": $AC \cap FG = E = (1, 1, 0)$, $AF \cap GC = B = (1, 0, 0)$, $AG \cap FC = D = (0, 1, 0)$.

## 1.5  Looking Ahead

Chapter 2 further develops properties of central collineations, including composition of central collineations and the important connection with Desargues' Theorem on triangles in perspective. The chapter provides extensive practice and challenges in carrying out central collineations. The exercises at the end of Chap. 2 include several classic problems for which central collineations lead to elegant proofs.

Chapter 3 is a selective survey of Euclid's *Elements*. The selection is to give the needed background in geometry for the reader. Some properties will be called on in Chaps. 5 and 7. Euclid's definition-postulate-proposition structure was followed in the *Conics* of Apollonius, a century later, and in much of mathematical presentation ever since.

Chapter 4 is an introduction to the study of conic sections from a historical viewpoint, centered on Greek geometer Apollonius, from around 200 BC. It also introduces the essential concept of the harmonic division of a line. Although ancient mathematics was without projective methods, an examination of conic sections from that era is a good starting place for a tour of projective geometry. While a modern text on projective geometry is likely to devote only a chapter or two, usually late in the book, to conic sections, the situation was quite different before the nineteenth century. Then, projective methods hardly existed outside the study of conic sections. It is important to see the Greek background on which projective methods would be applied from the seventeenth century on.

Chapters 5 and 6 move the story from the ancient Greeks to the seventeenth century and the beginnings of the projective study of the conic sections. We examine the work of Philippe de la Hire. He used central collineations to develop the study of conic sections in works of 1673 and 1685, built on the concept of a harmonic set of points. In Chap. 5, we recount what La Hire learned about tangents and the harmonic division of secants that cut a circle, culminating in the *pole-polar* relation of a point and line with respect to a given circle. In Chap. 6, properties of the circle that is the base of a cone carry in projection to the section of that cone by a plane.

Chapter 7 switches from the historical viewpoint to introduce an alternative definition of a harmonic set, by a *complete quadrilateral*. Specialized central collineations are introduced, including the *involution collineation*. Treating a conic section as the image of a circle under a central colleation, the dual theorems of Pascal and Brianchon appear here.

Chapter 8 is an examination of early nineteenth century projective geometry, when it became a subject in its own right, independent of the study of conic sections. Lazare Carnot's geometry of 1803 is noted, but most of the chapter is devoted to the work of Jean-Victor Poncelet. One thread through the work of Poncelet is his discovery of central collineations.

The foci of a conic, the subject of Chap. 9, have an interesting history in projective geometry. They are not projective in that they are not preserved by a central collineation. However, they are related in a way observed by Poncelet and further developed by Michel Chasles, a generation later.

Chapter 10 is an examination of Jacob Steiner's 1832 work, which changed the nature of projective geometry. Poncelet had drawn attention to projective transformations as a subject deserving study in its own right, and labeled as *homologous* the corresponding features of figures related by a projective transformation. Steiner sharpened the focus. He wrote of *projective forms*—lines and pencils of lines—and asked the crucial question: given a one-to-one correspondence of the lines or points of two projective forms, how do we know if they are projectively related, i.e., is the pairing the result of a projective transformation? His answer was that the forms are projectively related exactly when the cross-ratio of any four points equals that of the corresponding four points. Further, he gave a new definition of a conic section: the set of points of intersection of the pairs of two projectively related pencils. (A *pencil* is the set of lines on a particular point.)

This introduction to projective geometry concentrates on three figures: Philippe de la Hire, Jean-Victor Poncelet, and Jacob Steiner. Another figure looms over all this history, Girard Desargues. He wrote one major, although short, work on projective geometry and conic sections, in 1639. Then that work disappeared for two centuries. His work was confusing, introducing new concepts and new vocabulary, but when carefully studied it displays a depth of understanding unmatched for most of those two centuries. In Chap. 11, we take a limited look at some of the work of Girard Desargues.

Chapter 12 briefly views projective geometry after Steiner. Through this long history we see movement away from dependence on the concept of length. The complete quadrilateral allowed a definition without length of a harmonic set. Starting with K. G. C. von Staudt, in 1847, a group of geometers pushed even farther toward a geometry that depends only on projective properties, not length.

Chapter 13 is on matrix methods. It is widely known that collineations of the projective plane, represented by homogeneous coordinates, are performed through multiplication by non-singular 3-by-3 matrices. We take up the question of which matrices give central collineations, and then find a matrix that carries out a given central collineation and find the central collineation for a given matrix. With Chap. 1, it can be studied independently of the rest of this book.

In Chap. 14 we discuss three historically important applications of projective geometry: the area of a parabolic segment, by Archimedes; the construction of sundials by Ibrahim ibn Sinan (909–946); and application of central collineations in rules for perspective drawing, as developed by Brook Taylor. Isaac Newton's application of properties of conic sections in his *Principia Mathematica* is the subject of Chap. 15.

The most familiar transformations of the real plane, $\mathbf{R}^2$, and its extension to the real projective plane, are central collineations. These include reflection over a line, vertical stretch, and dilation. Rotation and translation are composition of two central collineations. The broad set of projective transformations are the compositions of central collineations, although, like Steiner, we think of a projective transformation as one which preserves the projective properties, such as the cross-ratio of four points. In this sense, central collineations are a unifying concept for a broad range of transformations.

## 1.6  Notes and Exercises

**Note** Any college library will have a shelf of books on projective geometry. A good one which includes coverage of central collineations is Pedoe's *An Introduction to Projective Geometry* [5]. An elementary introduction is in Courant and Robbin's *What is Mathematics?* [3, p 165–217]. A classic which develops projective geometry by means of central collineations is *Elements of Projective Geometry*, 1893, by Luigi Cremona [4]. This 1893 text is the second edition of an 1885 translation of the earlier Italian version, but has been augmented, particularly by a chapter on foci. This new material on foci had appeared in Michel Chasles's *Traité des Sections Coniques*, of 1865 [2], which, in turn, builds on material in Chasles's *Traité de Géométrie Supérieure*, of 1852 [1].

**Exercise 1** Here are the duals of Lemmas 1–4. Their proofs are left for the reader.

**Lemma 1.1′** *If a collineation has two centers, then it is the identity map.*

**Lemma 1.2′** *If a collineation $\phi$ has a center, A, then any fixed line not on A must be a line of fixed points.*

**Lemma 1.3′** *If a collineation $\phi$ has a center, A, and m is not a fixed line, then point $m \cap m'$ is a fixed point.*

**Lemma 1.4′** *If a collineation $\phi$ has a center, A, then it has a line of fixed points.*

**Exercise 2** By Properties $P1.$, $P2.$ and $P3.$, show that a projective plane must have at least seven points and seven lines.

**Exercise 3** Prove that if a projective plane has a finite number of points and Properties $P1.$, $P2.$ and $P3.$ hold, then all lines have the same number of points.

**Exercise 4** Suppose Properties $P1.$, $P2.$ and $P3.$ hold. State and prove the dual of Property $P3$.

**Exercise 5** Build an affine geometry by using ordered pairs of the numbers in $F_3$, listing the points of the affine plane, and equations of the lines. Then represent in homogeneous coordinates—triples—the points and lines already created. Then add the minumum number of points to a have a projective geometry.

How many points are on the line at infinity of the projective plane? How many lines are in the projective plane? (This question is largely answered in the Excursion at the end of Chap. 13.)

**Exercise 6** Suppose a geometry is given by its set of points and the grouping of those points into lines in such a way that Properties $P1.$, $P2.$ hold and that there are four points $A, B, C, D$ where no three are collinear. Prove that every line, $m$, has at least three points.

Suggestion: Consider Case 1, where line $m$ includes at most one of the points $A, B, C, D$, and then Case 2, where $m$ includes two points from $A, B, C, D$.

## 1.7  Some Hints and Solutions to Exercises

### Exercise 1
Lemma 1.1′. If a collineation has two centers, then it is the identity map.

Hint: Consider what it means to show that a mapping, $f$, is the identity: Take an arbitrary point, $X$, and show that $f(X) = X$.

Lemmas 1.2′–1.4′ can be proved by creating the duals of the solutions given in the chapter.

### Exercise 2
By Properties $P1.$, $P2.$ and $P3.$, show that a projective plane must have at least seven points and seven lines.

We have seen that the Fano Seven-Point Geometry satisfies Properties $P1.$, $P2.$ and $P3.$ There can be no fewer points or lines: By $P3.$, there are at least four points, $A, B, C, D$ with no three collinear. By $P1.$ there are the six lines $AB, AC, AD, BC, BD, CD$. By $P2.$, each pair of opposite sides meet in a new point: $AB \cap CD$, $AC \cap BD$, $AD \cap BC$. So there are at least seven points. And by $P1.$ and $P2.$, there is at least one new line on two or three of those points at which the opposite sides meet.

**Exercise 3**

Let $r$ and $s$ be different lines in a finite projective geometry. Say $r$ has $n$ points: $R_1, R_2, \ldots, R_n$ where $R_1 = r \cap s$, and $s$ has, at least, points $R_1$, $S_2$, and $S_3$. We'll show that $s$ has at least $n$ points. Let lines $R_2 S_2$ and $R_3 S_3$ meet at point $K$, which must lie on neither $r$ nor $s$. Then the lines $K R_2$, $K R_3, \ldots, K R_n$ meet $s$ in $n - 1$ points, in addition to $R_1$. So $s$ has at least $n$ points. We can reverse the roles of $r$ and $s$ to show that $s$ has no more than $n$ points.

# References

1. Michel Chasles, *Traité de Géométrie Supérieure*, Paris: Bachelier, 1852.
2. Michel Chasles, *Traité des Sections Coniques*, Paris: Gauthiers-Villars, 1865.
3. Richard Courant and Herbert Robbins, *What is Mathematics?*, New York and London: Oxford University Press, 1941; reissued, revised by Ian Stewart, 1996.
4. Luigi Cremona, *Elements of Projective Geometry* 2nd edition, transl. by C. Leudesdorf, Oxford: Clarenden Press, 1893.
5. Daniel Pedoe, *An Introduction to Projective Geometry*, Pergamon Press: New York, 1963.
6. Jan Vredeman de Vries [Friese], Perspective, in Samuel Marolois, *Opera mathematica, ou Oeuvres mathématiques tractans de géomémetrie, perspective, architecture et fortification*, Amsterdam: Ian Ianssen, 1651.

# Chapter 2
# Central Collineations: Properties

## 2.1 Specifying a Central Collineation

We recall that a collineation is a one-to-one onto mapping of the projective plane to itself in which collinear points are mapped to collinear points. The collineation is *central* if there is a *center*, a point $A$ where all lines on $A$ are fixed, meaning the line is mapped to itself, although individual points on the line need not be fixed. We have seen that a collineation is central exactly when it has a line of fixed points, an *axis*.

We will speak of a central collineation with center $V$ and *axis*, or line of fixed points, $l$, as a $(V, l)$ *collineation*, terminology and notation used in [3].

**Theorem 2.1 (Central Collineation Construction)** *A $(V, l)$ collineation is completely defined if we assign to one point $B$, not on $V$ or $l$, its image $B'$, provided $V$, $B$, and $B'$ are collinear.*

*Proof of Theorem 2.1* See Fig. 2.1, where $V$, $l$, $B$, and $B'$ are indicated. We first find the image of a point $Q$ which is not on line $BB'$. Draw line $QB$, meeting $l$ at fixed point $K$. $Q'$ must lie on line $KB'$, and $Q'$ must lie on line $VQ$. So $Q'$ is determined. Once we have found images for points not on line $BB'$, we can use these other points and their images to find images for points on $BB'$. We note that first knowing $Q$ and $Q'$ and $B$ would let us correctly find $B'$, so we get the same mapping of the plane when we are given any appropriate pair $B$ and $B'$. □

**Theorem 2.2** *Problem: For a given $(V, l)$ collineation, mapping $B$ to $B'$, find its vanishing line.*

*Solution for Theorem 2.2* See Fig. 2.2. We'll take a line $m$ on center $V$. We review how we find the image, $Q'$, of a point $Q$ on $m$, and from that process we'll decipher how we find the point, $E$, on $m$ that is mapped to infinity. We know that the axis and the vanishing line must be parallel since they can only meet at infinity, so one point on the vanishing line indicates the location of that line. We can see that $Q'$ was

© Springer Nature Switzerland AG 2020
C. Baltus, *Collineations and Conic Sections*,
https://doi.org/10.1007/978-3-030-46287-1_2

**Fig. 2.1** Construction of $Q'$

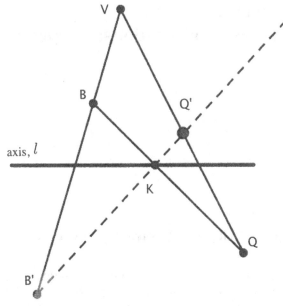

**Fig. 2.2** Constructing vanishing line

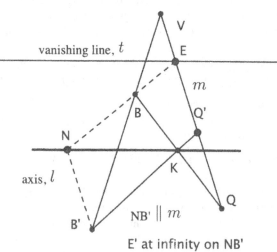

found, in the last step, by drawing $B'K$. So to find $E'$ at infinity, we need to find a point $N$ on $l$ so $NB' \parallel m$. Then $E$ will lie on $NB$.

**Theorem 2.3 (Construction from Vanishing Line)** *Problem: For a given $(V, l)$ collineation with vanishing line $t$, find the image of a point $Q$.*

*Solution for Theorem 2.3* See Fig. 2.3. We need to find two lines mapped onto $Q'$. One is $VQ$. For a second line mapped to $Q'$, we select a point $J$ on the vanishing line, $t$, and draw $JQ$, meeting the axis at a point, $K$. The image of line $JQ$ will lie

**Fig. 2.3** Find image when
vanishing line is given

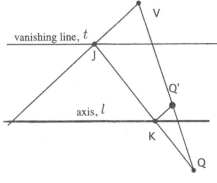

**Fig. 2.4** Theorem 2.4

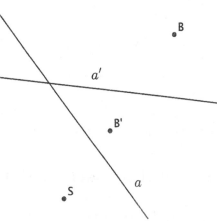

on $K$ and on $J'$, and on $Q'$. $J'$ will be at infinity in the direction of $VJ$. So we draw
on $K$ the line parallel to $JV$. That parallel line is $KQ'$, where $Q'$ is on $VQ$.

The Construction from Vanishing Line is the construction that Philippe de la Hire
introduced in the *Plani-coniques* section added to his 1673 [2]. We will see his work
in Chap. 6.

A central collineation may be defined in yet other ways. Here is an example.

**Theorem 2.4** *Given lines $a$ and $a'$ and point $S$ on neither $a$ nor $a'$, and given a
pair of points $B$ and $B'$ collinear with $S$, not collinear with $a \cap a'$, neither $B$ nor $B'$
on $a$ or $a'$, then there is a (unique) central collineation with center $S$ mapping $a$ to
$a'$ and $B$ to $B'$. See Fig. 2.4.*

We can prove the theorem by finding the axis. As we will see in Exercise 5, the
axis must lie on any point, except the center, at which a line and its image meet. So
$a \cap a' = P$ lies on the axis. Completing the solution is Exercise 19.

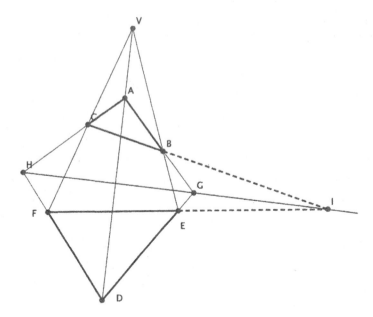

**Fig. 2.5**  Proof of Desargues' Theorem

## 2.2   Central Collineations and Desargues' Theorem

Desargues' Theorem and the existence of central collineations are closely related. In the real projective plane, for any given line $l$ and point $V$, and points $X$ and $X'$ collinear with $V$ in that plane, there is a unique central collineation with that center and axis, and which maps $X$ to $X'$. This is enough to prove Desargues' Theorem, which first appeared in 1648.

**Theorem 2.5 (Desargues' Theorem)**  *Suppose that $\triangle ABC$ and $\triangle DEF$ are in perspective from point $V$, i.e., the three lines $AD, BE, CF$ are concurrent at $V$. Then the corresponding sides of the two triangles meet in three collinear points.*

**Proof**  See Fig. 2.5. Let $AB$ and $DE$ meet in $G$ and let $AC$ and $DF$ meet in $H$. Let $\phi$ be the central collineation with center $V$ and axis $GH$ that maps $A$ to $D$. Then line $HA$ is mapped to line $HD$, so $C$, on line $HA$, must be mapped to a point on line $HD$. So $C'$ is $VC \cap HD$, which means $C' = F$. In the same way, we see that $B$ is mapped to $E$. Therefore, line $CB$ is mapped to line $FE$. A line and its image meet in a fixed point (Proof is an exercise), so $CB$ and $FE$ meet on the axis $GH$. This is what we needed to prove.                                                                      □

On the other hand, if Desargues' Theorem holds in a projective plane, then for any line $l$, point $V$, and points $A$ and $D$ collinear with $V$, we claim there is a $(V, l)$ central collineation $\phi$ that maps $A$ to $D$. Thus, this claim and Desargues' Theorem are equivalent in a projective plane. This is the reason that Desargues' Theorem may be added to the required properties of a projective plane, together with Properties

**Fig. 2.6** Proof that a
collineation is constructed

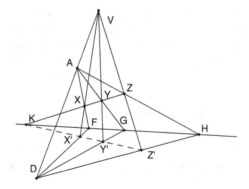

*P1., P2., P3.* We define the function $\phi$ in accord with the procedure given above. The pertinent claim is this theorem:

**Theorem 2.6** *In a projective plane, we take a line $l$, point $V$, and points $A$ and $D$ collinear with $V$. We define a function $\phi$ for, initially, any $B$ not on line $AV$ as follows: let $AB$ meet $l$ in point $G$. We define $B' = \phi(B)$ to be the point $VB \cap GD$. And $\phi$ is then defined for points on $VA$ following this procedure using $B$ and $B'$. And we assume Desargues' Theorem. Then $\phi$ is a collineation.*

***Proof for Collinear Points Not on*** *$AV$* See Fig. 2.6. Let $X$, $Y$, and $Z$ be collinear points not on line $AV$. We plot their images $X'$, $Y'$, and $Z'$, where $AX$, $AY$, and $AZ$ meet $l$ in points $F$, $G$, $H$, respectively. We need to show that $X'$, $Y'$, and $Z'$ are collinear. Let line $XYZ$ meet $l$ in point $K$. Now $\triangle AXY$ and $\triangle DX'Y'$ are in perspective from $V$, so by Desargues' Theorem their corresponding sides meet in collinear points: $K$, $F$, $G$, since $FG \cap XY = K$. So $X'$, $Y'$ and $K$ are collinear. Likewise, $\triangle AXZ$ and $\triangle DX'Z'$ are in perspective from $V$, so we conclude in the same way that $X'$, $Z'$ and $K$ are collinear. Together, this means that $X'$, $Y'$ and $Z'$ are collinear.                                                                                   □

## 2.3   Composition of Central Collineations

Although the composition of two collineations must be a collineation, when those collineations are central collineations their composition need not be a central collineation. But if, for example, the two central collineations have a common axis, $l$, their composition must be a central collineation with axis $l$. The placement of the center is considered in the following theorem.

**Theorem 2.7 (Central Collineation Composition Theorem)** *Let $\phi_1$ and $\phi_2$ be central collineations with centers $S_1$ and $S_2$, respectively, both with axis $l$. Then the composition $\phi_1 \circ \phi_2$ is a central collineation that has its center collinear with $S_1$ and $S_2$.*

**Proof** $\phi_1 \circ \phi_2$ is a central collineation since it has a line of fixed points, $l$. Where is the center? Let $m$ be the line on $S_1$ and $S_2$. $\phi_1 \circ \phi_2(S_2) = \phi_1(S_2)$, which is a point on $m$ since $m$ is on the center, $S_1$, of $\phi_1$. In a similar way, we see $\phi_1 \circ \phi_2(S_1)$ is on $m$. So $m$ is a fixed line of $\phi_1 \circ \phi_2$, so the center of $\phi_1 \circ \phi_2$ lies on $m$.                                                          $\square$

## 2.4   Group Properties

The set of central collineations with a common axis or the set with a common center, or which have both a common center and axis, forms a *group* under composition. Exercises 14 and 15 are proofs.

**Definitions III** A set of functions, $S$, forms a *group* under the operation of composition, $\circ$, if

(a)  the set is *closed* under $\circ$, i.e., $(f, \ g \in S) \rightarrow (f \circ g \in S)$,
(b)  the set is *associative* under $\circ$, i.e., $(f \circ g) \circ h = f \circ (g \circ h)$ for all $f, \ g, \ h \in S$,
(c)  there is an *identity* function, $e$, in $S$, so $e \circ f = f = f \circ e$ for all $f \in S$, and
(d)  for each $f \in S$ there is an *inverse* function, $f^{-1}$ in $S$, so $f \circ f^{-1} = e = f^{-1} \circ f$.

The group is *commutative* if, in addition, $f \circ g = g \circ f$ for all $f, \ g$ in $S$.

It helps to note that function composition is always associative. For functions $f, g, h$, $((f \circ g) \circ h)(x) = (f \circ g)(h(x)) = f(g(h(x)))$, and we get the same expression for $(f \circ (g \circ h))(x)$.

## 2.5   Excursion: Two Commutative Groups of Central Collineations

**Definitions IV** An *elation* is a central collineation whose center lies on its axis.
A *homology* is a central collineation whose center is not on its axis.

It follows from the Central Collineation Composition Theorem that the set of elations with a common axis, $l$, forms a group under function composition. If elations $\phi_1$ and $\phi_2$ have distinct centers $V_1$ and $V_2$ on a common axis, $l$, then the composition $\phi_3 = \phi_2 \circ \phi_1$ also has axis $l$ and the center, $V_3$, of $\phi_3$ lies on $l$. The set of elations with both a common axis and common center will form a group.

What is not so clear is that such groups are commutative.

**Theorem 2.8** *Let elations* $\phi_1$ *and* $\phi_2$ *have centers* $V_1$ *and* $V_2$, *respectively, and have a common axis,* $l$, *where* $V_1 \neq V_2$. *Then* $\phi_1 \circ \phi_2 = \phi_2 \circ \phi_1$.

**Theorem 2.9** *Let elations* $\phi_1$ *and* $\phi_2$ *have the same center,* $V$, *and a common axis,* $l$. *Then* $\phi_1 \circ \phi_2 = \phi_2 \circ \phi_1$.

**Fig. 2.7** Elations with a
common axis are
commutative

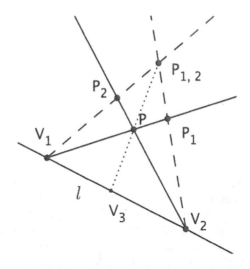

***Proof of Theorem 2.8*** (This proof and the next follow [3, p 106].) Let $P$ be a point
not on $l$, and let $P_1 = \phi_1(P)$ and $P_2 = \phi_2(P)$. Let $P_{1,2} = \phi_1(P_2)$ and $P_{2,1} = \phi_2(P_1)$. Now (see Fig. 2.7) $P_{1,2}$ lies on line $V_1 P_2$ and on line $\phi_1(V_2 P)$. This latter
line is $V_2 P_1$, so $P_{1,2}$ is the intersection of $V_1 P_2$ and $V_2 P_1$. Analogous reasoning tells
us that $P_{2,1}$ is the same point $P_{1,2}$. Since $\phi_1 \circ \phi_2$ and $\phi_2 \circ \phi_1$ map a point $P$ not on
$l$ to the same image, and have a common axis and common center $l \cap P P_{1,2}$ then
$\phi_1 \circ \phi_2 = \phi_2 \circ \phi_1$. □

We note that when $V_1 \neq V_2$ then $V_3$, the center of $\phi_1 \circ \phi_2$, equals neither $V_1$ nor
$V_2$. For if, say, $V_1 = V_3$, then for a line $m$ on $V_1$, $\phi_1(m) = \phi_3(m) = m = \phi_1 \cdot \phi_2(m)$,
so $\phi_2(m) = m$, which is not possible when $V_1 \neq V_2$.

***Proof of Theorem 2.9*** Let $\alpha$ be an elation with axis $l$ but center $V_1 \neq V$. We know
$\phi_2 \circ \alpha = \alpha \circ \phi_2$, and that this transformation is an elation with center on $l$ but not
equal to $V$. So $\phi_1 \circ (\phi_2 \circ \alpha) = (\phi_2 \circ \alpha) \circ \phi_1 = \phi_2 \circ (\alpha \circ \phi_1) = \phi_2 \circ (\phi_1 \circ \alpha)$. $\alpha$ has
an inverse, from which we deduce that $\phi_1 \circ \phi_2 = \phi_2 \circ \phi_1$. □

Another commutative group is the set of $(V, l)$ collineations for a given axis $l$ and
a given center $V$ not on $l$. Exercise 23 asks the reader to illustrate the commutativity
by a construction in a particular case. Chapter 10 Exercise 7 asks for a proof, which
is provided in the Solution section of Chap. 10.

## 2.6   Notes and Exercises

**Note 1**  Central collineation constructions that utilize the vanishing line require the
construction of parallels. Parallels can be constructed by just a straightedge if certain
figures are given. An interesting case is found in G.J. 'sGravesande's [1, Tab. 3
*Fig.* 3, 1727] (or from the 1777 edition), in which a given parallelogram enables us
to construct by straightedge alone a parallel to a given line on a given outside point.

**Fig. 2.8** 'sGravesande
construction, 1727

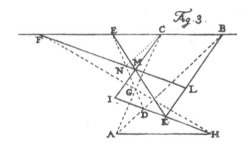

See Fig. 2.8. 'sGravesande justified the construction by the elementary Euclidean geometry involving parallels and triangle similarity. Here is the construction and the corresponding justification utilizing a central collineation called a *dilation*. *Given*: parallelogram $IKLM$, line $BC$, and point $A$. We are to construct on $A$ a parallel to $BC$ by straightedge only. Let parallelogram sides and one diagonal, $KL$, $IM$, $MK$, and $ML$, meet line $BC$ at $B$, $C$, $E$, and $F$, respectively. Draw $AB$, meeting $IK$ at $D$. Draw $DE$ and $AC$, meeting at $G$. Let $FG$ meet $IK$ at $H$. Then $AH \parallel BC$.

***Proof*** We make use of the dilation, or central collineation, with center $G$, mapping $D$ to $N = LM \cap DE$, so the line at infinity is the axis. Since any line is mapped to a parallel line, then line $ADB$ is mapped to the line on $N$ that is parallel to $ADB$.

We then use Euclid's Book VI Prop 2 (Side Splitter Theorem—see Chap. 3, Thm. 3.1) that a line parallel to one side of a triangle cuts the other two sides in proportion, and conversely:

$$\frac{EC}{CB} = \frac{EM}{MK} = \frac{EN}{ND}.$$

Thus $NC$ is the line on $N$ that is parallel to $ADB$. So our dilation, which maps $H$ to $F$, also maps $A$ to $C$, and line $AH$ to line $FC$. So $AH \parallel BC$. (Poncelet gave the same construction in *Cahier* 7 of 1813, p 437.)                              □

Note 2 Pedoe's [3, p 29–32] gives a most interesting example, credited to F.R. Moulton, of a projective plane in which Desargues' Theorem does not hold. The plane satisfies Properties *P1.* and *P2.*, namely that any two lines determine exactly one point, and that any two points determine exactly one line. And there is an abundance of points and lines. This serves to emphasize the importance of Desargues' Theorem when we work in a projective plane other than the real projective plane.

**Exercise 1** State and prove the dual of the *Central Collineation Construction Theorem*.

**Exercise 2** The dual of Desargues' Theorem is its converse. Desargues' Theorem can be used to prove its converse. A proof is not hard to find, as in [3, p 27]. Proofs

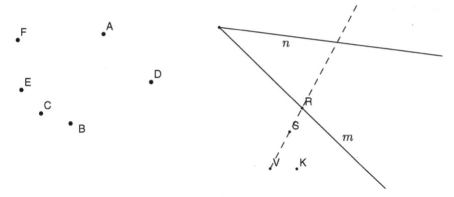

**Fig. 2.9** Chapter 2 Exercises 8, 9

of Desargues' Theorem in 3-dimensions are also easy to find, as in [3, p 28]. Use a central collineation to prove the dual of Desargues' Theorem.

**Exercise 3** Suppose $\phi$ is a $(V, l)$ central collineation, $V$ not on $l$, that is not the identity. Show that any fixed point other than $V$ must lie on $l$.

**Exercise 4** Suppose $\phi$ is a $(V, l)$ central collineation where $V$ lies on $l$. If there is a fixed point $K$ not on $l$, show $\phi$ must be the identity.

**Exercise 5** Under a central collineation, show that a line (not on the center) and its image meet in a fixed point, and that that fixed point must lie on the axis of the central collineation.

**Exercise 6** Suppose $\phi$ is a central collineation and that a line $m$ not on center $V$ is a fixed line. Prove that $m$ is the axis.

**Exercise 7** Suppose $\phi$ is a $(V, l)$ central collineation and that points $R$ and $S$ are fixed points which are collinear with $V$. Prove that line $RS$ is a line of fixed points.

**Exercise 8** Mark six points: $A$, $B$, $C$, $D$, $E$, $F$, with no three collinear and let $\phi$ be a central collineation where $E$ is fixed, $A$ is mapped to $B$, and $C$ is mapped to $D$. Construct the image of $F$. See Fig. 2.9 Left.

**Exercise 9** Suppose, as in Fig. 2.9 Right, $\phi$ is the central collineation with center $V$, mapping $R$ to $S$, and where the line of fixed points is the line at infinity. $R$ lies on $m$.

*(a)* Find the image of line $m$.
*(b)* Find the image of point $K$.
*(c)* Find the image of line $n$.

**Exercise 10** Show that reflection over a given line is a central collineation. What is the center? Note that a rotation is the composition of two reflections, so we will

**Fig. 2.10**  Exercise 11

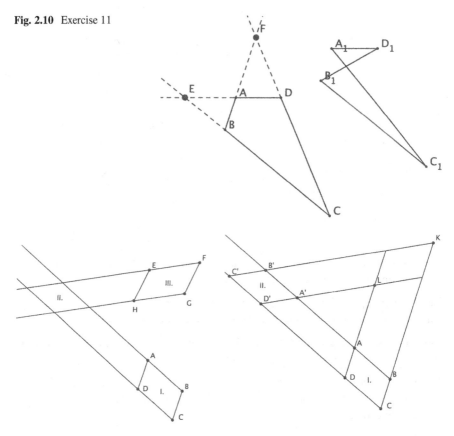

**Fig. 2.11**  Exercise 12

have shown that rotation is the composition of central collineations. Is translation a central collineation?

**Exercise 11**  See Fig. 2.10. Show that any quadrilateral with no parallel sides can be mapped to a parallelogram by a central collineation. Where should the vanishing line be placed? Show the same thing for a quadrilateral with exactly one pair of opposite sides parallel. Note in the diagram that four points can be joined in several ways to form a quadrilateral.

**Exercise 12**  See Fig. 2.11. To show that any given Parallelogram I can be mapped in two central collineations to a given Parallelogram III, the plan is to create Parallelogram II as pictured, and then give a central collineation that maps Parallelogram I, $ABCD$, to Parallelogram II. A similar mapping would send Parallelogram II to the given Parallelogram III. With Fig. 2.11 Right, to map Parallelogram I to Parallelogram II, here is a suggestion: Use a central collineation that maps the point at infinity $AD \cap BC$ to point at infinity $A'D' \cap B'C'$. Now, give the center, $V$. Then give the axis, $l$, so $A \rightarrow A'$ and $B \rightarrow B'$. Explain why Parallelogram $ABCD$ is mapped to Parallelogram $A'B'C'D'$.

**Exercise 13**  We can create a composition of central collineations, applying Exercises 11 and 12, to create a collineation mapping any given quadrilateral to a given parallelogram, and, further, to create a collineation mapping any given quadrilateral to any other given quadrilateral. In fact, there is a composition of central collineations mapping any four points, no three collinear, to any other set of four points, no three collinear. Explain, please.

**Exercise 14**  For a given line, $l$, and a given center, $V$, prove that the set of $(V, l)$ collineations forms a group under function composition.

**Exercise 15**  For a given line, $l$, prove that the set of central collineations with axis $l$ forms a group under function composition.

**Exercise 16**  Classic problem [6, Problem 27]. See Fig. 2.12 Upper Left. Given triangles $ABC$, $A'B'C'$ and $A''B''C''$, where $BC$, $B'C'$, $B''C''$ meet at $D$, $AB$, $A'B'$, $A''B''$ meet at $E$, and $AC$, $A'C'$, $A''C''$ meet at $F$, with $D$, $E$, $F$ collinear. We know by Desargues' Theorem that $AA'$, $BB'$, $CC'$ are concurrent at a point, $G$, that $AA''$, $BB''$, $CC''$ are concurrent at a point $J$, and $A'A''$, $B'B''$, $C'C''$ are concurrent at a point $K$.

Show $G$, $J$, $K$ are collinear. Suggestion: Consider two central collineations with axis $DEF$ and their composition.

**Exercise 17**  Classic Problem [4, p 18], from a Cambridge Scholarship Examination. Given quadrilateral $ABDC$, with $AB \cap CD = U$, $AC \cap BD = V$, $BC \cap UV = G$, $AD \cap UV = F$, and $BF \cap AC = L$. Prove $FC$, $LG$, and $AB$ are concurrent. See Fig. 2.12 Upper Right.

**Exercise 18**  Classic Problem [4, p 10]. Given a triangle $XYZ$ and a point $P$. Draw $PX$, $PY$, $PZ$, meeting sides $YZ$, $XZ$, $XY$ at $F$, $G$, $H$, respectively. Then draw $GH$, meeting $XZ$ at $L$, draw $FH$, meeting $XZ$ at $M$, and draw $FG$, meeting $XY$ at $N$. Prove $L$, $M$, $N$ are collinear. (Line $LMN$ is called the *polar* of $P$ with respect to the triangle $XYZ$.) See Fig. 2.12 Center Right.

**Exercise 19**  In Fig. 2.12 Lower Left, construct the axis $l$ so the resulting $(S, l)$ collineation maps line $a$ to $a'$ and point $B$ to $B'$.

**Exercise 20**  State the dual of Theorem 2.4 and illustrate it.

**Exercise 21**  Jacob Steiner [5], Article 21, 1832, Part I with his Fig. 27. See our Fig. 2.12. Let lines $a$, $a_1$, $a_2$ be concurrent at $E$. From point $B$, line $a$ is in perspective with line $a_1$, with point $A$ paired with $A_1$ and with point $C$ paired with $C_1$; from point $B_1$, line $a$ is in perspective with line $a_2$, with point $A$ paired with $A_2$ and with point $C$ paired with $C_2$; and from point $B_2$, line $a_1$ is in perspective with line $a_2$, with point $A_1$ paired with $A_2$ and with point $C_1$ paired with $C_2$. Prove that $B$, $B_1$, and $B_2$ are collinear. Steiner used this lemma to prove Desargues' Theorem;

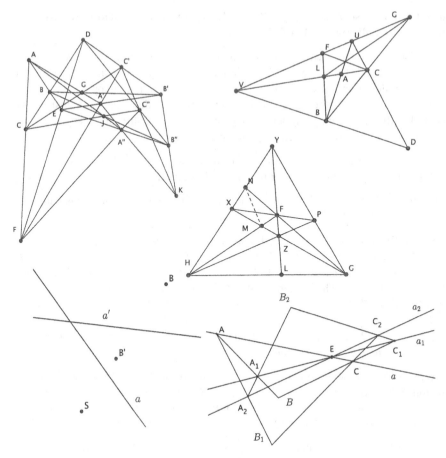

**Fig. 2.12** Upper Left: Exercise 16, Upper Right: Exercise 17. Center Right: Exercise 18, Lower Left: Exercise 19. Lower Right: Exercise 21, from Steiner's *Fig.* 27 of 1832 Art 21 Part I

see if you can prove it without calling on Desargues' Theorem. Suggestion: Central Collination Composition Theorem.

**Exercise 22** Jacob Steiner [5], Article 21 Part III, 1832. See Fig. 2.13. Let lines $a$, $a_1$, $a_2$ be concurrent at $E$. Fix points $B$ and $B_1$ on none of the three lines. Then take any triangle with vertices $A$, $A_1$, $A_2$ on lines, respectively, $a$, $a_1$, $a_2$, and side $AA_2$ on $B$ and side $AA_1$ on $B_1$. Then there is a point $B_2$ collinear with $B$ and $B_1$ so that no matter which triangle is chosen in this way, side $A_1A_2$ lies on $B_2$.

**Exercise 23** Let $\phi_A$ and $\phi_B$ be $(V, l)$ collineations with a common axis $l$ and a common center $V$. Illustrate by a careful diagram that on some line $n$, not $l$ and not on $V$, $\phi_A \circ \phi_B = \phi_B \circ \phi_A$. Suggestion: See Fig. 2.14. Draw collinear $V$, $B$, $B'$ and collinear $V$, $A$, $A'$ where $V$, $A$ and $B$ are not collinear, $\phi_B(B) = B'$, etc., and let

**Fig. 2.13** Exercise 22, from
Steiner's 1832 Art 21 Part III
Exercise 19

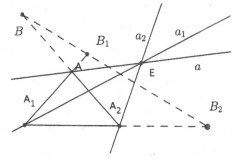

**Fig. 2.14** Exercise 23

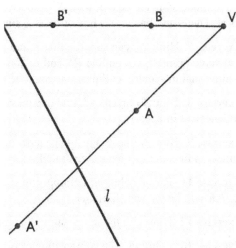

$n$ be line $AB$. Construct $\phi_A \circ \phi_B(n)$ and $\phi_B \circ \phi_A(n)$. (We revisit this problem in
Chap. 10, Exercise 7.)

## 2.7   Some Hints and Solutions

**Exercise 3** Solution: We'll suppose that $P$ is a fixed point of $\phi$ and that $P$ is not
on axis $l$. Then $\phi$ is the identity. Why? Let $X$ be a point that is not on $VP$ and does
not lie on $l$. Let line $PX$ meet $l$ at $Z$. Since $P$ and $Z$ are fixed, then $X'$ lies on line
$PX$. Further, $X'$ lies on line $VX$ since $V$ is the center. $PX \cap VX = X$ so $X' = X$.
We then would have a collineation with at least two fixed points off the line of fixed
points, so $\phi$ would be the identity.

**Exercise 4** Solution: First, let $X$ be a point on neither $l$ nor line $VK$. Let $KX$ meet
the axis, $l$, at point $Z$. Line $KX$ is a fixed line under $\phi$ since $K$ and $Z$ are fixed, so $X'$
lies on $KX$. Likewise, $X'$ lies on line $VX$ since $V$ is the center. So $X' = VX \cap KX$,

which equals $X$. Second if $Y$ is a point on $VK$, we can repeat this argument using $X$ as the fixed point to show that $Y' = Y$.

**Exercise 5**  Solution: Let $V$ be the center and $l$ the axis of the central collineation. Let $m$ be a line not on $V$ and $m'$ its image under the central collineation, with $m \cap m' = K$. $K'$ in on $m'$ since $K$ is on $m$. But, also, $K'$ is on line $VK$ since $V$ is the center. $m' \cap VK$ is $K$, so point $K$ is fixed. If $V$ lies on $l$ and the central collineation is not the identity, then by Exercise 4, $K$ lies on $l$. When $V$ is not on $l$, fixed point $P$ must lie on $l$ since otherwise the collineation would be the identity.

**Exercise 6**  Solution: Let $X$ be a point on fixed line $m$. Then $X'$ is on $m$ since $m$ is fixed. Since $V$ is the center, then $X'$ also lies on line $VX$. $VX \cap m = X$ so $X$ is fixed. Therefore, line $m$ is a line of fixed points.

**Exercise 7**  Hint: We can suppose that $R$ and $S$ do not both lie on axis $l$. First show that some point $P$, not on line $RS$ and not on axis $l$, is a fixed point, from which it follows that the central collineation would be the identity.

**Exercise 8**  Hint: To find axis $l$, find one fixed point in addition to $E$, say where line $AC$ meets it image.

**Exercise 9**  Hint: A line and its image meet on the axis. When the axis is the line at infinity, a line and its image are parallel.

**Exercise 11**  Partial solution: Let opposite sides of the quadrilateral meet at $F$ and $E$. $FE$ will be the vanishing line.

**Exercise 12**  Hint: Use the intersection of parallel lines $AB$ and $CD$ as $V$.

**Exercise 14**  Solution: The composition of two collineations is a collineation since any line will be mapped to a line. When the two collineations are $(V, l)$ central collineations, the composition will map fixed point $V$ to itself and each point of line $l$ to itself. A collineation with a line of fixed points, $l$, is a central collineation, and $V$ must be the center of that composition of collineations. This shows closure.

The inverse of the $(V, l)$ collineation that maps a point $A$ to $A'$ is the $(V, l)$ collineation, with the same $V$ and $l$, that maps $A'$ to $A$.

Function composition is always associative, the identity map is a $(V, l)$ collineation.

**Exercise 16**  Solution: Let $\phi_1$ and $\phi_2$ be central collineations with axis $DEF$; let $\phi_1$ have center $G$ and map $A$ to $A'$, while $\phi_2$ has center $J$ and maps $A''$ to $A$. Since $F$ is fixed and $A \rightarrow A'$ by $\phi_1$ , then $C \rightarrow C'$, and then $B \rightarrow B'$. In a similar way, $C'' \rightarrow C$, and then $B'' \rightarrow B$ by $\phi_2$. Then $\phi_1 \circ \phi_2$ maps $A''$ to $A'$, $B''$ to $B'$, and $C''$ to $C'$. So the center of $\phi_1 \circ \phi_2$ is $K$. By the Central Collineation Composition Theorem, $G$, $J$, and $K$ are collinear.

**Exercise 17**  Solution: Define $X$ to be $AB \cap FC$. Define a central collineation to have center $X$, axis $BD$, and where $A \rightarrow U$. Since $D$ is fixed and $A \rightarrow U$, then $F \rightarrow C$. $L'$ must lie on $BG$ since $B$ is fixed and $F \rightarrow C$. Further, $L'$ lies on $UV$ since $V$ is fixed and $A \rightarrow U$. So $L'$ is $G$. Therefore, line $LG$ lies on the center, $X$.

**Exercise 18** Solution: Consider the central collineation with center $P$ and axis $MN$, mapping $X$ to $F$. Since $N$ is fixed and $X \rightarrow F$, then $Y \rightarrow G$. As $M$ is fixed and $X \rightarrow F$, then $Z \rightarrow H$. This means line $YZ$ is mapped to $GH$. A line and its image meet in a fixed point, so $L$ is fixed. Other than $P$, any fixed point is on the axis, so $L$ is on axis $MN$.

**Exercise 19** Solution: Draw a line on $S$, not on $B$, meeting lines $a$ and $a'$, at $K$ and $K'$, respectively. Then line $BK$ will be mapped to line $B'K'$, and the intersection of these two lines, at a point $Q$, must also lie on the axis. We can check that the central collineation with center $S$, axis $PQ$, and mapping $B$ to $B'$, maps $a$ to $a'$.

**Exercise 21** Solution: Let $l$ (not on the diagram) be a line on $E$ which is on none of the other named points. Let $\phi$ be the central collineation with center $B$, axis $l$ and mapping $A_1$ to $A$, and let $\phi_1$ be the central collineation with center $B_1$, axis $l$ and mapping $A$ to $A_2$. Then, by the Central Collination Composition Theorem, $\phi_1 \circ \phi$ is a central collineation with axis $l$ and center collinear with $B$ and $B_1$. Now $\phi_1 \circ \phi$ maps $A_1$ to $A_2$ and $C_1$ to $C_2$, so its center is $B_2$.

**Exercise 22** Solution: Let $l$ (not in the diagram) be a line on $E$ which is on none of the other named points. Let $\phi$ be the central collineation with center $B$, axis $l$ and mapping $A_2$ to $A$, and let $\phi_1$ be the central collineation with center $B_1$, axis $l$ and mapping $A$ to $A_1$. Then, by the Central Collination Composition Theorem, $\phi_1 \circ \phi$ is a central collineation with axis $l$ and its center, which we label $B_2$, collinear with $B$ and $B_1$. Since $\phi_1 \circ \phi$ maps $A_2$ to $A_1$, then triangle side $A_1 A_2$ lies on $B_2$.

# References

1. G. J. 'sGravesande, *Matheseos universalis elementa*, Leiden: Samuel Luchtmans, 1727.
2. Philippe de La Hire, *Nouvelle Méthode en Géométrie pour les Sections des Superficies coniques et Cylindriques*, Paris, 1673.
3. Daniel Pedoe, *An Introduction to Projective Geometry*, Pergamon Press: New York, 1963.
4. E. A. Maxwell, *Methods of Plane Projective Geometry based on the use of General Homogeneous Coordinates*, Cambridge University Press, 1946, 1963.
5. Jacob Steiner, *Systematische Entwicklung der Abhägigkeit geometrischer Gestalten*, Erster Theil, Berlin: G. Fincke, 1832, in *Jacob's Steiner's Gesammelte Werke*, Erster Band, edited by K. Weierstrass, Berlin: G. Reimer, 1881, 229–460.
6. I. M. Yaglom, *Geometric Transformations III*, transl. A. Shenitzer, New Mathematical Library Vol. 24, Random House: New York, 1973.

# Chapter 3
# The Geometry of Euclid's *Elements*

## 3.1 Ancient Greek Mathematics Before Euclid

To start at the beginning, as with many stories in mathematics, we go back to ancient Greece, and the "Greek miracle." There were several Greek miracles, modern features that appeared as from nowhere. They may be related. There was the appearance of democracy, in Athens, around 500 BC. Around the same time philosophy emerged, reasoning by evidence and logical deduction about the great questions: what is our world made of, how and why does it change, how should we live? And mathematics that was formal and deductive.

In computational mathematics, in its application to practical problems and in texts for instructing students in the art, Greece was a millennium behind its neighbors in Mesopotamia and Egypt. This earlier mathematics could not be without reasoning. But that reasoning was not discussed in the fragments that have survived, suggesting that reasoning in mathematics was not formal and that exposition of that reasoning was not expected. The Babylonian tablet BM 13901—BM indicating the British Museum—poses the problem "I summed the area and two-thirds of my square-side and it was 0;35," where the 0;35 is 35/60. The text continues with the computation to find the side of the square: one-half [4, p 24]. The statement of the question indicates it was a question about area, not algebra. The "square-side" refers to a strip whose width is 1 and whose length is the side of the square, and the two-thirds of my square-side can indicate two strips of width 1/3. The solution can be read as a manipulation of areas. See Fig. 3.1. Reasoning, yes, but not formal.

Greeks introduced formal reasoning in geometry. In the remote past, on the border between myth and history, there was Thales of Miletus, active around 600 BC. It is reported that Thales stated and proved theorems. Among these theorems is that base angles of an isosceles triangle are equal. We do not know what argument he gave [3, vol 1, p 130].

We should note the fragmentary character of our information on Greek mathematics, especially before Euclid's *Elements*, which is from about 300 BC. Our

© Springer Nature Switzerland AG 2020
C. Baltus, *Collineations and Conic Sections*,
https://doi.org/10.1007/978-3-030-46287-1_3

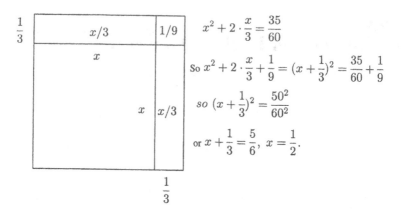

**Fig. 3.1**  Solution of Babylonian tablet problem BM 13901

most important source is by Proclus, a commentator of the fifth century AD. His *Commentary on Euclid* included a six-page summary of early Greek geometry. His information came, in turn, largely from a lost *History* by Eudemus, apparently written a couple decades before Euclid, but to which was added information on Euclid's *Elements*. A further source is Eutocius of Askalon, from the sixth century AD, who reports a history by Eratosthenes (third century BC) of construction problems [5, p 4]. Further information can be gleaned from works of Plato (c. 429–347 BC), Aristotle (384–322 BC), and even the plays of Aristophanes (c. 446–386 BC).

By the late fifth century BC, there were texts on geometry, in which new theorems were proved based on previously proved theorems. The first person who can confidently be called a mathematician was Hippocrates of Chios, producing a geometry around 440 BC [6, p 275].

Early in the fourth century BC, Plato, who did some mathematics himself, characterized mathematics and mathematical reasoning as essential in the education of a ruler. He wrote soon after the Peloponnesian War had left Greece in ruins. In an uncertain world, mathematics gave an example of certainty and reasoning to certainty. Thus the Greek miracles of philosophy and mathematics may be connected. And democracy, as it survived, deeply involved public argumentation. Those who studied and taught this public argumentation were called Sophists, from which comes the modern "sophistry." Contrasting with the tricks and strategizing of public argumentation, and the emerging philosophical skepticism, mathematics was, for Plato and others, the model of correctness and certainty.

Aristotle studied at Plato's *Academy*. He wrote books on logic that defined the subject into the early modern era. Included in his work was a discussion of the structure of a science: based on experience, we find indubitable truths, which he labeled *axioms* or *common notions*—common to all sciences—and *postulates*—particular to one science. So we have, for example, Euclid's first Postulate, that any two points can be joined by a line. From the axioms and postulates, we deduce propositions.

The *Elements* of Euclid, written in Alexandria, Egypt, about 300 BC, is the earliest comprehensive mathematics text which has survived into modern times. Proclus wrote that Euclid "brought together the Elements, systematizing many of the theorems of Eudoxus, perfecting many of those of Theatetus, and putting in irrefutable demonstrable form propositions that had been rather loosely established by his predecessors" [2]. It is written in thirteen "books." In structure, it illustrates Aristotle's model of a science. It is largely based on lost earlier work. It may be that Euclid's *Elements* alone survived the centuries of war and neglect because it was so well written and so widely studied. It served, with modification, as the primary introduction to geometry for students well into the twentieth century.

Here we give an abbreviated treatment of the geometry of Euclid's *Elements*, as background to Greek mathematics, and as background in plane geometry for the reader.

## 3.2 The Geometry of Euclid's *Elements*: A. Preliminaries in Book 1

Book 1 begins with *Definitions*. Then, following the form recommended by Aristotle, are the *Postulates* and *Common Notions*, followed by *Propositions*.

These, in modern language, are the definitions of most interest to us.

Definition 4. Lines which meet in congruent adjacent angles are called *perpendicular* and the angles thus formed are called *right angles*.

Definition 15. A circle is designated by its center and radius; it is the set of points whose distance from the center equals the given radius.

Definition 23. Lines which lie in one plane and never meet, even if produced indefinitely in both directions, are called *parallel*.

The first three postulates are assertions about straightedge/compass constructions.

Postulate 1. To draw a segment from any point to any other point.

Postulate 2. To extend a segment in a straight line.

Postulate 3. To describe a circle with any center and radius from that center.

In modern mathematics, these are widely reformulated as assertions about existence. In this sense, Eulid's First Postulate may now appear as "On any two points there is exactly one line." We will use "line" where Euclid used "straight line," and "segment" indicates the part of a line joining two points (with the endpoints).

After the postulates are the common notions. Two which we will have reason to cite are the first and fifth.

Common Notion 1. Things equal to the same thing are equal to each other.

Common Notion 5. The whole is greater than the part.

Here, modernized, are the foundational propositions of most interest.

Prop. 1. To draw an equilateral triangle on a given side.

Prop. 3. Given two unequal segments, we can cut from the longer a segment equal to the shorter.

Prop. 4. $SAS$ Triangle Congruence.

Props. 5 and 6. *Isosceles Triangle Theorem.* Two sides of a triangle are congruent exactly when the opposite angles are congruent.

Prop. 8. $SSS$ Triangle Congruence.

Prop. 26(part). $ASA$ Triangle Congruence.

In modern school vocabulary, line segments and angles have real number *measures*, and when two segments or angles agree in measure they are called *congruent*. *Equal* is reserved for numbers, such as segment lengths, or two names for the same figure.

**Definitions V** Two triangles are *congruent*, with a given correspondence of vertices, if the three pairs of corresponding sides and three pairs of corresponding angles are congruent.

In Euclid, and most geometry written since then, congruent figures are simply called *equal*. We simplify the matter by letting, for example, $AB$ represent both the segment, as a set of points, and its length, depending on the context. Likewise, $\angle X$ will represent both the angle as a set of points and the measure of the angle.

Propositions 1 and 3 of Book 1 illustrate Euclid's approach. He developed a sequence of propositions, each as the consequence of the definitions, postulates, common notions, and earlier propositions. For Proposition 1, $AB$ is the given side. See Fig. 3.2 Top Left. Justified by Postulate 3, we draw a circle with center $A$ and radius $AB$, and a second circle, with center $B$ and radius $BA$. Let $C$ be a point at which these two circles meet. By Postulate 1, we draw segments $AC$ and $BC$. The lengths $AC$ and $AB$ are equal, as are $BC$ and $BA$, by the definition of a circle. $AC = BC$ by Common Notion 1. Thus we have an equilateral triangle.

That no postulate guarantees the existence of such a point $C$ has led to various repairs of Euclid's set of postulates, especially in the late nineteenth century.

Proposition 3 declares that given two unequal segments, we can cut from the longer, $CD$, a segment equal to the shorter, $AB$. The point of the proposition is that we can set the opening of a compass to length $AB$ and then move the compass with that set opening to mark off length $AB$ on segment $CD$.

How is Proposition 3 proved? See Fig. 3.2 Right. We draw segment $AC$ (Post. 1), then construct an equilateral triangle $ACE$ (Prop. 1). $EA$ is extended past $A$ and $F$ marked on this extension so $AF = AB$. Then we draw the circle with center $E$ and radius $EF$, meeting the extension of $EC$ at $H$. Finally we draw the circle with center $C$ and radius $CH$. Where that circle meets $CD$, at $G$, we have segment $CG$ equal to $AB$.

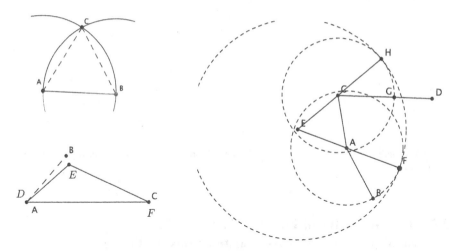

**Fig. 3.2** Diagrams, Euclid's *Elements* Book 1. Upper Left: Prop. 1, Right: Prop. 3, Lower Left: Prop. 4

Proposition 4, *SAS*, is the first of the triangle congruence propositions. The others that follow depend, eventually, on Proposition 4.

Euclid's proof of Proposition 4 is unusual in that it depends on the superposition of a figure, a move not justified by a postulate. Modern writers have, for the most part, opted to make *SAS* a postulate. When the given triangles are $\triangle ABC$ and $\triangle DEF$, with $\angle A \cong \angle D$, $AB = DE$, and $AC = DF$, then $\triangle ABC$ is superposed on $\triangle DEF$ with $A$ on $D$, side $AC$ on side $DF$, and with $B$ and $E$ on the same side of the line $DF$. Then $C$ lies on $F$ since $AC$ equals $DF$, and $B$ lies on $E$ since $\angle A \cong \angle D$ and $AB = DE$. Since the triangles coincide, they are equal. See Fig. 3.2 Lower Left.

We omit Euclid's lengthy proofs of Propositions 5 and 6, and leave a proof of Proposition 8, *SSS* Triangle Congruence, for the exercises.

The *ASA* Triangle Congruence Theorem is part of Prop. 26. It is justified by an indirect proof, also known as *proof by contradiction*. See Fig. 3.3 Left. We are given two triangles, $ABC$ and $DEF$, with $\angle B \cong \angle E$, $\angle ACB \cong \angle DFE$, and with congruent included sides $BC$ and $EF$. If we can show $AB = DE$, then it will follow by Prop. 4 that $\triangle ABC \cong \triangle DEF$. So we suppose $AB \neq DE$; say $AB > DE$. We mark point $G$ on $AB$ so $BG = ED$ (Prop. 3). Draw $GC$. Now by Prop. 4, $\triangle GBC \cong \triangle DEF$, so $\angle GCB \cong \angle DFE$. However, $\angle GCB$ is less in measure than $\angle ACB$, since by Common Notion 5 the whole is greater than its part, contradicting the given condition that $\angle ABC \cong \angle DEF$. Therefore, $AB$ and $DE$ are equal, from which we conclude that $\triangle ABC \cong \triangle DEF$.

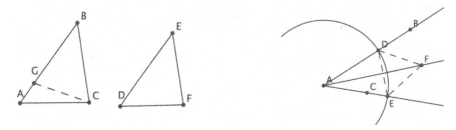

**Fig. 3.3** Diagrams, Euclid's *Elements* Book 1. Left: Prop. 26 ASA Congruence, Right: Prop. 9: Angle Bisector

## 3.3  The Geometry of Euclid's *Elements*: B. Straightedge/Compass Constructions in Book 1

Propositions 9 to 12, and 23, provide the procedure—and the justification—for these constructions: to bisect an angle, to bisect a segment, to construct a perpendicular to a given line on a given point, and to construct an angle congruent to a given angle. The justifications depend on the $SSS$ Triangle Congruence Theorem.

We illustrate Euclid's approach with  Proposition 9: To bisect a given angle $BAC$. See Fig. 3.3 Right. We take a point $D$ on side $AB$, and construct the circle with center $A$ that lies on $D$ (Post. 3). Let the circle meet side $AC$ at $E$. Construct equilateral triangle $DFE$ on Base $DE$ (Prop. 1). Join $A$ and $F$. $\triangle ADF \cong \triangle AEF$ by $SSS$ (Prop. 8). So corresponding angles $DAF$ and $EAF$ are congruent.

## 3.4  The Geometry of Euclid's *Elements*: C. Angles and Parallels

We now turn to a sequence of propositions concerning angles, parallel lines, and, in the pivotal Prop. 16, an exterior angle of a triangle.

Prop. 16. *Euclid's Exterior Angle Theorem*: An exterior angle of a triangle is greater in measure than either opposite interior angle. The shorter name *Exterior Angle Theorem* refers to Prop. 32: An exterior angle of a triangle is equal in measure to the sum of the two opposite interior angles.

There is a crucial foundational distinction between Propositions 16 and 32: Prop. 16 can be proved without invoking the *Parallel Postulate*, while the stronger Prop. 32 requires it. There is a corresponding distinction between Prop. 17, that any two angles of a triangle are together less than two right angles, and Prop. 32, that the three angles of a triangle are together equal to two right angles. And there is yet another corresponding distinction between Prop. 27, that two lines cut by a transversal so alternate interior angle are congruent must be parallel, and its converse, Prop. 29. In each pair, the first proposition can be proved without the Parallel Postulate. The geometry developed without any postulate about the existence of parallels is called *Absolute Geometry*. Euclid's Propositions 1 through 28 of Book 1 are part of Absolute Geometry. It was proved in the nineteenth century that the second proposition of each of these pairs requires the Parallel Postulate, or an equivalent assumption.

Here are the essential propositions through Prop. 28.

Prop. 13. *Linear Pair Theorem.* If angles $ABD$ and $CBD$ form a linear pair, i.e., they are adjacent with $B$ on line $AC$ and between $A$ and $C$, as in Fig. 3.4, then the angles are together equal to two right angles. (The converse is Proposition 14.)

Prop. 15. Vertical angles are equal. (See Fig. 3.4.)

Prop. 16. An exterior angle of a triangle is greater than either opposite interior angle.

Prop. 26. $ASA$ and $AAS$. If two angles and the included side, or two angles and a corresponding non-included side, are congruent in two triangles, then the triangles are congruent. (We have already seen the proof of $ASA$.)

Prop. 27 and 28. *Alternate Interior Angle Theorem.* If two lines are cut so alternate interior angles are congruent, then the lines are parallel. Likewise, if corresponding angles are congruent, then the lines are parallel.

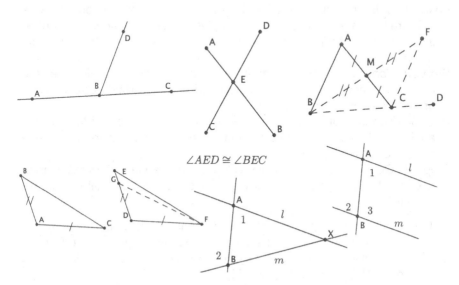

**Fig. 3.4** Diagrams, Euclid's *Elements* Book 1. Upper Left: Angles $ABD$ and $DBC$ are a Linear Pair. Upper Middle: Prop. 15, Vertical angles are congruent, Upper Right: Prop. 16. Lower Left: Prop. 26 $AAS$ Congruence, Lower Middle: Prop. 27. Lower Right: Prop. 29

For the proof of Proposition 16, see Fig. 3.4 Upper Left. With $\triangle ABC$, let side $BC$ be extended past $C$ to $D$, creating exterior angle $ACD$. Let $M$ be the midpoint of $AC$ ($M$ can be constructed by Prop. 10), then draw $BM$ (Post. 1) and extend it past $M$ (Post. 2), marking $F$ on this extension so $BM = MF$ (Prop. 3). Join $F$ and $C$ (Post. 1). Then $\angle BMA \cong \angle FMC$ (vertical angles). By $SAS$, $\triangle AMB \cong \triangle CMF$. So, as corresponding parts of congruent triangles (part of Prop. 4), $\angle A \cong \angle ACF$. $\angle ACD > \angle ACF$ (Common Notion 5). It follows that exterior angle $ACD$ is greater than opposite interior angle $A$.

Then the $AAS$ part of Proposition 26 is proved. See Fig. 3.4. We are given $\triangle ABC$ and $\triangle DEF$ where $\angle A \cong \angle D$, $\angle B \cong \angle E$, and $AC = DF$. The strategy is to show that $AB$ equals $DE$ by supposing that $AB < DE$ and deriving a contradiction.

Assuming $AB < DE$, mark $G$ on $DE$ so $DG = AB$ (Prop. 3). We draw $GF$, and note that $\triangle DGF \cong \triangle ABC$ ($SAS$). With, then, $\angle DGF \cong \angle B$, we have $\angle DGF \cong \angle E$, which violates Euclid's Exterior Angle Theorem. Therefore, $AB = DE$, so $\triangle ABC \cong \triangle DEF$ ($SAS$).

Here is the proof of the *Alternate Interior Angle Theorem*, Proposition 27.

We suppose $l$ and $m$ are cut at $A$ and $B$, respectively, so alternate interior angles 1 and 2 are equal. Further, we suppose $l$ and $m$ are not parallel, meeting at a point $X$. See Fig. 3.4. But then exterior angle 2 equals opposite interior angle 1 of $\triangle ABX$, violating Euclid's Exterior Angle Theorem. We conclude that $l$ and $m$ are parallel.

Beginning with Prop. 29, Euclid presented propositions whose proofs require the Parallel Postulate. We state the Parallel Postulate, Postulate Five, followed by several propositions that depend on it.

Postulate 5. If a line cuts two other lines so the interior angles on the same side are together less than two right angles, then the two lines, if produced indefinitely, meet on that side on which the angles are less than two right angles.

Prop. 29. *Converse of Alternate Interior Angle Theorem*. If parallel lines are cut by a transversal, then alternate interior angles (and corresponding angles) are congruent.

Prop. 32. The interior angles of a triangle are, together, equal to two right angles, and an exterior angle of a triangle equals, in measure, the sum of the two opposite interior angles.

Prop. 33. Opposite sides of a parallelogram are equal. (A parallelogram is defined to be a quadrilateral where both pairs of opposite sides are parallel.)

To prove Proposition 29, we suppose $l$ and $m$ are parallel, but are cut by line $AB$ so $\angle 2 > \angle 1$. Now, the sum $\angle 2 + \angle 3$ equals two right angles (Linear Pair), so $\angle 3 + \angle 1$ is less than two right angles. By Post. 5, $l$ and $m$ meet, contradicting what was supposed. Therefore, $\angle 2 \cong \angle 1$. See Fig. 3.4.

Book 1 ends with the Pythagorean Theorem and its converse, in Propositions 47 and 48.

## 3.5 The Geometry of Euclid's *Elements*: D. Triangle Similarity and Circles in Books 6 and 3

Although similar triangles are only introduced in Book 6, it will help to give a definition and several theorems from Book 6 which we will apply in considering Book 3, on circles.

**Definitions VI (Book 6, Def. 1)** Two polygons are *similar* if there is a correspondence of vertices so corresponding angles are equal and corresponding sides are in proportion.

Book 6 Propositions 4, 5, 6: Triangles are similar when there is a correspondence of vertices so that

$AA$: two pairs of corresponding angles are congruent, or
$SSS$: the three pairs of corresponding sides are in proportion, or
$SAS$: two pairs of corresponding sides are in proportion and the included angles congruent.

The first proposition of Book 6 tells us that the area of a triangle is one-half its base times its height. It follows that triangles with the same height and bases on the same line have areas in the ratio of those bases. We call Euclid's Book 6 Prop. 2 the *Side Splitter Theorem*, on which Euclid bases his triangle similarity theorems.

**Theorem 3.1 (Side Splitter Theorem)** *If a line parallel to one side of a triangle cuts the other two sides of the triangle, then it cuts those sides in proportion. And, conversely, a line that cuts two sides of a triangle in proportion must be parallel to the third side.*

Euclid's proof depends on area, a common strategy with both Euclid and Apollonius. See Fig. 3.5 Left. We use $(XYZ)$ to denote the area of $\triangle XYZ$. Let $DE$ be parallel to base $BC$ of $\triangle ABC$, where $E$ lies on side $AC$ an $D$ lies on side $AB$. Then $\triangle DEC$ and $\triangle DEB$ have equal areas because they have equal heights and share the base $DE$. Now, $\triangle AED$ and $\triangle ECD$, which have the same height, have areas in the ratio of their bases $AE$ and $EC$. Likewise, $\triangle AED$ and $\triangle BED$,

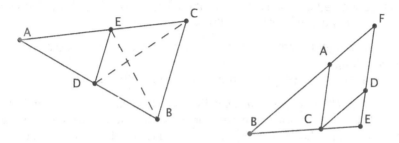

**Fig. 3.5** Euclid Book 6. Left: Prop. 2 *Side Splitter Theorem*. Right: Prop. 4, Proof of the $AA$ Similarity Theorem

which have the same height, have areas in the ratio of their bases, $AD$ and $DB$. So $AE : EC = (AED) : (ECD) = (AED) : (BED) = AD : DB$. The same concepts are used in the proof of the converse. □

In Book 6, we assume that when two angles of a triangle agree in measure, then the third angles are congruent.

**Theorem 3.2 (Book 6, Prop. 4, AA Triangle Similarity)** *If two angles of one triangle are congruent to two angles of a second triangle, then the triangles are similar with the given correspondence of vertices.*

**Proof** See Fig. 3.5 Right. We are given two triangles, $ABC$ and $DCE$, with congruent corresponding angles. We build the triangles so base $BC$ is continued in base $CE$, with $A$ and $D$ on the same side of that base. $ED$ and $BA$ are extended to meet at $F$. (That they meet is guaranteed by the Parallel Postulate.) By Prop. 28 of Book 1, $FD$ is parallel to $AC$, and $BF$ is parallel to $CD$. Since opposite sides of a parallelogram are congruent (Book 1 Prop. 33), $AF = CD$ and $AC = FD$.

By the Side-Splitter Theorem, $AB : AF = CB : CE$. By what was just shown, $AB : CD = CB : CE$. Likewise, since $BF$ is parallel to $CD$, then $BC : CE = FD : ED = AC : DE$. It follows that triangles $ABC$ and $DCE$ are similar.     □

The $SSS$ and $SAS$ similarity theorems are proved in the same fashion in Propositions 5 and 6 of Book 6.

We next move to Book 3 of the *Elements*, on circles. It develops propositions of importance in our examination of projective geometry. First, some terminology, followed by two important propositions.

**Definitions VII** A *tangent* is a line or segment that meets a circle in exactly one point, even if extended (Book 3 Def. 2). Prop. 17 is a construction of a tangent to a circle from an outside point.

These terms are not in Euclid:
A *chord* of a circle is a segment joining two points of a circle. A *diameter* is a chord that lies on the center of the circle. A *secant* is a line that meets a circle twice and extends outside the circle. An angle whose vertex is at the center of a circle is a *central angle*, and an angle whose vertex is on a circle, where each of whose sides again meets the circle, is an *inscribed angle*. An *arc* is a connected part of a circle, and the *measure* of an arc is the measure of the central angle that intercepts the arc.

Prop. 18. A tangent meets a circle perpendicular to the radius drawn to the
    point of tangency.
Prop. 19 is the converse, whose proof makes use of Prop. 18.

**Proof of Prop. 18 (A Variation on Euclid's Proof)** Let a tangent meet a circle $F$ at point $C$. If the tangent is not perpendicular to radius $FC$, then construct (Book 1 Prop. 12) the perpendicular to the tangent from $F$, meeting the tangent at $G$. In right triangle $GFC$, hypotenuse $FC$ is longer than leg $FG$, which means $G$ lies inside

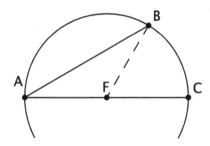

**Fig. 3.6** Euclid Book 3 Prop. 20 *Inscribed Angle Theorem*. Case when a side of the angle is a diameter. $arc(BC) = \angle BFC = 2 \cdot \angle BAC$. Note that the measure of exterior angle $BFC$ is the sum of the measures of the congruent angles $A$ and $B$

the circle. But $G$ is on the tangent so it cannot lie inside the circle. We conclude that the tangent at $C$ is perpendicular to radius $FC$.

Here are further theorems of importance.

Proposition 20. *Inscribed Angle Theorem.* The measure of an inscribed angle is half the measure of the intercepted arc. (See Fig. 3.6.)

Book 3 Propositions 35 and 36 develop what we call, following Jacob Steiner in the nineteenth century, the *Power-of-a-Point*, $X$, with respect to a given circle. See Fig. 3.7, Center for Prop. 35 and Left for Prop. 36.

Prop 35. If chords $AB$ and $CD$ meet at $X$, then the rectangle with sides $AX$ and $BX$ equals (in area) the rectangle with sides $CX$ and $DX$, i.e., $AX \cdot BX = CX \cdot DX$.

Prop. 36. If point $X$ is outside the circle, and a secant on $X$ meets the circle at $A$ and $B$ while a tangent from $X$ meets the circle at $C$, then $AX \cdot BX = CX^2$.

Corollary to Proposition 36 (Not in Euclid): The two tangent segments drawn to a circle from an outside point are equal in length.

We can summarize in a theorem.

**Theorem 3.3 (Power-of-a-Point)** *Given a circle and a point $X$, then for any secant or chord on $X$ meeting the circle at $A$ and $B$, the value $AX \cdot BX$ depends only on $X$ and the given circle, independent of the secant or chord $AB$.*

Modern proofs of Propositions 35 and 36 are most often based on triangle similarity. Euclid's proofs are different, for he developed the triangle similarity theorems only in Book 6.

Here is a modern proof of Euclid's Book 3, Propositions 35 and 36.

***Proof of Prop. 35*** See Fig. 3.7 Center. Let $X$ be inside the circle, with chords $AB$ and $CD$ meeting at $X$. The claim is that $AX \cdot XB = CX \cdot XD$. We draw $AC$ and $BD$. By the Inscribed Angle Theorem, Euclid's Book 3 Prop. 20, $\angle ACD \cong \angle ABD$ and $\angle CAB \cong \angle CDB$, since the angles of each pair intercept the same arc. So $\triangle ACX \sim \triangle DBX$ by AA, so $\dfrac{AX}{XD} = \dfrac{CX}{XB}$ and the theorem follows.

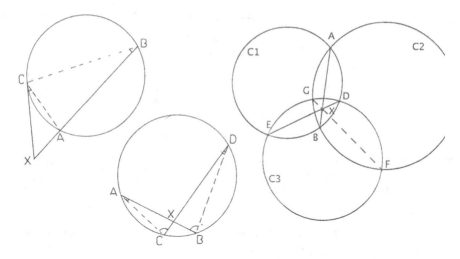

**Fig. 3.7** Left and Center: Power-of-a-Point $X$, Right: Common Secant Theorem

In the case handled in Prop. 36, with $X$ be outside the circle, we use the same idea, with $\triangle XAC \sim \triangle XCB$. See Fig. 3.7 Left. We note( Euclid's Book 3 Prop 32) that $\angle XCA$ has half the measure of the intercepted arc, $CA$.

An immediate consequence of the Power-of-a-Point Theorem is the following

**Theorem 3.4 (Common Secant Theorem. (Carnot) [1, 1803 Art. 305])** *Let three circles have, pairwise, three common secants. Then those common secants are concurrent.*

***Proof*** See Fig. 3.7 Right. Let the three circles be $C_1$, $C_2$, $C_3$, where $C_1$ and $C_2$ meet in points $A$ and $B$, circles $C_1$ and $C_3$ meet in points $D$ and $E$, and $C_2$ and $C_3$ meet in $F$ and $G$. Let $AB$ meet $DE$ in point $X$. By the Power-of-a-Point Theorem applied to circle $C_1$, $XA \cdot XB = XD \cdot XE$. Now, this means the "power" of point $X$ is the same with respect to all three circles. Let $k$ denote that power. Now suppose the line on $X$ and $F$ meets circle $C_2$ at point $G_2$, and meets circle $C_3$ at point $G_3$. By the power of point $X$ applied to circles $C_2$ and $C3$, lengths $XG_2$ and $XG_3$ both equal $k/XF$, so $XG_2 = XG_3$. This is only possible if $G_2 = G_3 = G$. Thus, the three secants are concurrent at $X$.                                                                    □

## 3.6   Exercises

**Exercise 1** Suppose a diameter of a circle meets a chord, not a diameter. Show that the diameter is perpendicular to the chord if and only if it bisects the chord.

**Exercise 2** Prove the $SSS$ Congruence Theorem:

*Given* $\triangle ABC$ and $\triangle DEF$ where $AB = DE$, $AC = DF$ and $BC = EF$.
*Prove* $\triangle ABC \cong \triangle DEF$.

Idea: Place the triangles so $A$ and $D$ coincide, so $B$ and $E$ coincide, and $C$ and $F$ lie on opposite sides of line $AB$. Draw line $CF$. (Take the case where lines $CF$ and $AB$ meet between $A$ and $B$. We can also assume that in the case of adjacent angles $WXY$ and $YXZ$ whose interiors do not overlap, $m\angle WXY + m\angle YXZ = m\angle WXZ$.)

**Exercise 3** Prove Euclid's Book 3 Prop. 19: If a line meets a circle at a point $C$ perpendicular to the radius drawn to point $C$, then the line is tangent at $C$ to the circle.

Hint: Suppose the line meets the circle at $C$ perpendicular to the radius to $C$ and, further, meets the circle at a second point, $D$. Explain why the resulting figure violates at least one theorem.

# References

1. Lazare Carnot, *Géométrie de Position*, Paris: Duprat, An XI, 1803.
2. Euclid, *Euclid's Elements*, edited and translated by David Joyce, https://mathcs.clarku.edu/~djoyce/java/elements/elements.html, 1998.
3. Sir Thomas Heath, *A History of Greek Mathematics*, Oxford: Clarendon Press, 1921; reissued by Dover, New York, 1981.
4. Victor Katz, *A History of Mathematics: An Introduction* 3rd edition, Boston: Addison-Wesley, 2009.
5. Wilbur Richard Knorr, *The Ancient Tradition of Geometric Problems*, Boston: Birkhauser, 1986.
6. Reviel Netz, *The Shaping of Deduction in Greek Mathematics: A Study in Cognitive History*, Cambridge: Cambridge University Press, 1999.

# Chapter 4
# Conics in Greek Geometry: Apollonius, Harmonic Division, and Later Greek Geometry

## 4.1 Conic Sections in Ancient Greece

Why would Greek geometers study the conic sections? We have no direct answer to that question in any ancient text that has survived. It was not, generally, for the scientific applications of conic sections. The earliest practical applications we know of are in sundials, with medieval Arabic scientists. Only in the early modern period, with Galileo (1564–1642), was the parabolic path of bodies in motion recognized. And only with Kepler (1571–1630) were the orbits of planets seen to be ellipses. An exception to this picture is a short work, *On Burning Mirrors*, whose author, Diocles, was active about 200 BC. It was, it seems, the first to give the focus of a parabola, with the property that a surface formed by revolving a parabola about its axis would reflect the sun's rays to the focus of that parabola [7, p. 124]. But it did not spark work in applications; there is no surviving mention of *On Burning Mirrors* until the sixth century AD, by Eutocius of Askalon, and, again, in Islamic work of the eleventh century.

For the ancient Greeks, at least two factors are worth mentioning. First, Greek geometers prized simplicity. In the plane, the simplest and most fundamental shapes are the line and the circle. That they could be constructed was the content of Postulates 1 and 3 of Euclid's *Elements*, and those constructions are the bases of further constructions in the *Elements*. When we move to three dimensions, we see in Euclid the same preference for the simplest shapes: the sphere, pyramid, and cone, and solids with regular polygons as faces. The cone, in the third century BC, was generated by a line on a fixed point in space running along the circumference of given circle; its volume is given in Euclid's Book XII, Proposition 10.

Second, the conic sections could find solutions to problems. Consider the three famous problems of Greek antiquity: to trisect a given angle, to construct the edge of a cube whose volume is double that of a given cube, and to construct a square whose area equals that of a given circle. These seem to have been already well known in the time of Hippocrates. "The Birds," a play by Aristophanes, of 414 BC,

© Springer Nature Switzerland AG 2020
C. Baltus, *Collineations and Conic Sections*,
https://doi.org/10.1007/978-3-030-46287-1_4

has a playful reference to squaring the circle. The character Meton, an astronomer and mathematician of the time, was described as making a construction "in order that your circle may become square." [6, vol. 1, p. 220]

Hippocrates himself is credited with transforming the problem of doubling the cube. Proclus wrote, "They say the first to effect the reduction [of doubling the cube to finding two mean proportionals] of the difficult constructions was Hippocrates of Chios." [10, p. 253] Hippocrates recognized that the cube of edge $a$ could be doubled if $x$ and $y$ could be found so that

$$\frac{a}{x} = \frac{x}{y} = \frac{y}{2a}.$$

$x$ and $y$ are called *mean proportionals* of $a$ and $2a$. Over the decades, various procedures were developed to find mean proportionals, including one attributed to Plato [6, vol 1, pp. 255–257]. Expressed in the nomenclature of Pappus, mean proportionals $x$ and $y$ could be found by *solid loci*. The *solid loci* are the conic sections, and it seems that the likely discoverer of the conic sections, Menaechmus, a contemporary of Aristotle, was the same person who applied these curves to finding the two mean proportionals. As described by Eutocius, translated into modern notation, with parabolas $x^2 = ay$ and $y^2 = 2ax$ and hyperbola $xy = 2a^2$, the first two meet at point $(x, y)$ where $\frac{a}{x} = \frac{x}{y} = \frac{y}{2a}$, and the second and third curves meet at the same point [10, pp. 278–284].

Thus, the conic sections may have been discovered as a means to solve a problem in mathematics. We'll see below how Apollonius arrived at these equations a century after Menaechmus, suggesting how Menaechmus might have approached the problem himself. (Menaechmus did not use the names "parabola" and "hyperbola.")

Within a half century after Menaechmus, the elementary properties of the conic sections were known, but works by Euclid and Aristaeus developing those properties have been lost. Only the *Conics*, by Apollonius of Perga, from about 200 BC, survived, just barely [4, p. 7]. From Pappus's *Mathematical Collection*, of the early fourth century AD, we know something about the lost works by Euclid and Aristaeus. Further, Archimedes made reference to conic properties as if the properties were widely known; Archimedes worked between the generations of Euclid and of Apollonius.

## 4.2   The *Conics* of Apollonius

The *Conics* of Apollonius was the essential classic, studied by all who studied conic sections into the early modern era, written in eight books late in the third century BC. As with Euclid, we know almost nothing about Apollonius. He was from Perga, in modern-day Turkey, and worked at Alexandria, the center of learning at the time.

And as with Euclid, much of the content of the *Conics* is based on earlier work that is now lost.

The *conic section*, for Apollonius and all mathematicians into the nineteenth century, was the section—or intersection—$ELD$, of a conic surface by a slicing plane. See Fig. 4.1 Left. His fundamental definition is of a *diameter*.

Apollonius began his work with definitions, and so shall we.

**Definitions VIII**   A *conic surface* is that traced by a line, of indefinite length in both directions, which stays on a given point, the *vertex*, in space as the line runs along a circle (in a *base plane*).

A *cone* is that part of a conic surface between the base circle and the vertex.

A *conic section*, or just *conic*, is the intersection of the conic surface with a slicing plane that does not meet the vertex and which meets the base plane in a line $FG$.

(When the slicing plane is parallel to the base plane, the section is a circle, which may explain why Apollonius did not include the circle among the possible conic sections.)

An *axial triangle*, $ABC$, is a triangle whose base is a diameter $BC$ of the base circle and whose vertex is the vertex, $A$, of the cone.

When the slicing plane meets some axial triangle twice (in addition to the base), the conic section produced is an *ellipse*; when the slicing plane meets one side of an axial triangle and meets a second side extended (neither side is the base), then the conic section is a *hyperbola*; otherwise—when the slicing plane meets just one side of an axial triangle—the conic section is a *parabola*.

Suppose in a conic section a straight line bisects all the chords which are parallel to some particular chord. Then that straight line is a *diameter*, and each of the bisected chords is a pair of *ordinates*, one on each side of the diameter (or the entire chord may be called the *ordinate*).

When a diameter of a conic is an ordinate for another diameter of the conic, those diameters are called *conjugate diameters*.

A point at which a diameter meets the conic section is a *vertex* of the conic (with respect to that diameter). When a diameter meets a conic section in two vertices, the midpoint of those vertices is a *center* of the conic section. It is a theorem that an ellipse and a hyperbola have just one center, while a parabola has none.

Apollonius then proved that any plane slicing a conic surface meets an axial triangle in a diameter. In particular,

**Theorem 4.1**   *Suppose a conic surface with vertex A is sliced by a plane that meets the base plane in line FG and meets the conic surface in conic ELD. Draw the diameter of the base circle which is perpendicular to FG, meeting the base circle in B and C. Let ED be the intersection of the slicing plane with axial triangle ABC. Then ED is a diameter of conic ELD.*

**Proof**   See Fig. 4.1 Left, which illustrates the case where $FG$ is outside the base circle, making the conic section an ellipse. The cone is again sliced, but this time parallel to the base plane and through a point $M$ on $ED$, giving the circle $PLR$,

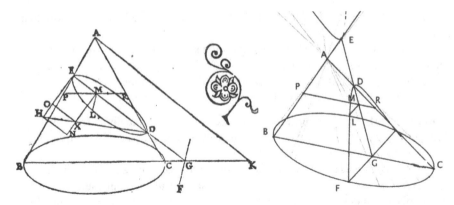

**Fig. 4.1** Left: Ellipse, Commandino's *Conics* of Apollonius, 1696 ed. Right: Hyperbola $FDRE$ as slice of a conic surface

where $L$ is on the ellipse and $P$ and $R$ are on the axial triangle. Now $PR \perp LM$, so continuing $LM$ gives a chord of the ellipse whose midpoint is $M$. (A diameter bisects any perpendicular chord in a circle). Since the circle parallel to the base plane is arbitrary, any such circle gives a chord of the ellipse parallel to the base plane and perpendicular to a line parallel to $PR$, then $DE$ bisects all those parallel chords. Thus $ED$ is a *diameter*, and those chords are *ordinates*.        □

To derive the equation of the ellipse, note that by the Power-of-a-Point Theorem, for point $M$, $ML^2 = PM \cdot MR$. By similar triangles,

$$ML^2 = PM \cdot MR = EM \cdot MD \cdot \frac{BG \cdot CG}{EG \cdot DG} = EM \cdot MD \cdot Constant1. \quad (4.1)$$

Apollonius defined the *parameter* or "erect side" (*latus rectum* in Latin) as the segment $EH$ that is perpendicular to $ED$ at $E$ and whose length is $Constant1 \cdot ED$, which we denote as $p$.

See Figs. 4.1 Left and 4.2 Center. Apollonius set the *latus rectum* perpendicular to the slicing plane, but it works just as well if, following Claude Mydorge [8], 1631, we place the *latus rectum* in the slicing plane.

When line $HD$ meets, at $X$, the line on $M$ parallel to $EH$, then by similar triangles $\dfrac{MX}{MD} = \dfrac{EH}{ED}$. Then Eq. (4.1) becomes

$$ML^2 = EM \cdot MX = EM \cdot MD \cdot \frac{p}{ED}. \quad (4.2)$$

From Fig. 4.1 Right, we derive the abscissa-ordinate equation of the hyperbola. In this case we can assume that the slicing plane meets the base plane in a line $FG$ that meets the base circle twice. We get exactly the same similar triangles, from which follows the same equation

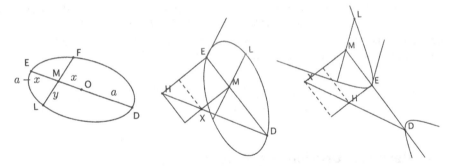

**Fig. 4.2** Left: Modern equation of ellipse. Center and Right: Latus Rectum $EH$ for ellipse and hyperbola, drawn in the plane of the conic

$$ML^2 = PM \cdot MR = EM \cdot MD \cdot Constant1 = EM \cdot MD \cdot \frac{p}{ED}. \qquad (4.3)$$

Again, $Constant1 = (BG \cdot CG)/(EG \cdot DG)$ and $Constant1 \cdot ED$ is the length $p$ of the *parameter* or *latus rectum*, represented in Fig. 4.2 as segment $EH$ in the plane of the section, perpendicular to $ED$. Again, $MX$ is parallel to $EH$, with length that satisfies $\dfrac{MX}{MD} = \dfrac{EH}{ED}$, from which we arrive at

$$ML^2 = EM \cdot MX. \qquad (4.4)$$

The diameter $ED$, found this way, is now called the "principal diameter." It turns out there is nothing special about this "principal diameter." In fact, Apollonius spent much of Book 1 showing that any other chord on the center of an ellipse or hyperbola is also a diameter, with an abscissa-ordinate equation of the same form, but with a different parameter.

We also have an important theorem:

**Theorem 4.2** *Let $ED$ be a diameter of a conic, where $E$ is a vertex. Then the tangent at $E$ to the conic is parallel to the ordinates of diameter $ED$. [Apollonius Book I Props. 17, 32]*

**Proof (*Not the Proof of Apollonius*)** If the line on $E$ and parallel to the ordinates should meet the conic in another point $K$, then diameter $ED$ would not bisect all the chords in a family of parallel chords, meaning it violates the definition of a diameter. □

When we look at $MX$ in the plane of the conic section, we see that with the ellipse, $MX$ is shorter than the latus rectum $EH$, and the area of rectangle $HX$ is subtracted from that of rectangle $HM$ to give area $EM \cdot MX$. In the case of the hyperbola, the areas are added to give area $EM \cdot MX$. This is the origin of the names of the sections: *ellipse* from the Greek for "falls short" and *hyperbola* from the Greek for "exceeds."

See Fig. 4.2 Left. Note that if we let $2a$ be length $ED$ and $x$ the directed distance from the center, $O$, to $M$, and let $ML = y$, then $EM \cdot MD = (a - x)(a + x)$ and the equation is $y^2 = (a^2 - x^2) \cdot Constant1$, from which the familiar equation of the ellipse follows (when $a^2 \cdot Constant1 = b^2$):

$$\frac{x^2}{a^2} + \frac{y^2}{b^2} = 1. \qquad (4.5)$$

## 4.3   Harmonic Division of a Segment

A central concept in the study of conic sections by Apollonius was the harmonic division of a segment associated with a conic. Early modern writers, especially Desargues and La Hire, would give the concept even more emphasis. Apollonius did not give a definition and he did not point out the common appearance of the harmonic division with the three conic sections. That he did not treat the circle as a conic section may have made it harder to treat harmonic division as a common property of the conics.

**Definitions IX**  If we have four collinear points, in order $A - B - C - D$, we say $A$ and $C$ are **harmonic conjugates** of $B$ and $D$, a relation which we represent as $H(AC, BD)$, provided

$$AD \cdot BC = AB \cdot CD, \text{ with exactly one of } B \text{ and } D \text{ between } A \text{ and } C. \qquad (4.6)$$

When the order is clear, we simply say $A, B, C, D$ is a *harmonic set*. Note that

$$H(AC, BD) \text{ iff } H(BD, AC)$$

and that interchanging $A$ and $C$, or $B$ and $D$, preserves the harmonic relationship.

Note that three collinear points and a given order uniquely define the fourth point of a harmonic set. More precisely,

**Theorem 4.3**  *When point $D$, collinear with $A$ and $C$, is not between $A$ and $C$, then there is a unique point $B$ between $A$ and $C$ where $H(AC, BD)$.*
**Proof**  When $AB = x$, then equation

$$\frac{AB}{BC} = \frac{AD}{CD} \text{ is } \frac{x}{AC - x} = K$$

for which there is a unique solution. When the order of the points is $A - C - D$, we can check that $0 < x < AC$.                                             □

The following will also be important.

**Theorem 4.4**  *Equivalent to Eq. (4.6) is*

$$AM^2 = MB \cdot MD \tag{4.7}$$

*when M is the midpoint of AC.*

**Proof**  First note that if the order of points is $A-B-C-D$ and $AD \cdot BC = AB \cdot CD$, then $AB > BC$, so the order is $A-M-B-C-D$. Then $AD \cdot BC = AB \cdot CD$ is $(2AM + CD)(MC - MB) = (AM + MB)CD$. Expanding and simplifying, with $AM = MC$, we get $2AM^2 - 2AM \cdot MB - CD \cdot MB = MB \cdot CD$, or

$2AM^2 = 2AM \cdot MB + 2CD \cdot MB$. Since $AM + CD = MD$, the equation follows.                                                                                    □

## 4.4  Conics and the Harmonic Relation

Apollonius showed in Book 1 Props 17, 32 and 37, and in Book 3 Prop 37, how the harmonic division of a line, as La Hire would put it, appeared with the conic sections; tangents to a conic are part of this connection.

See Fig. 4.3. Let a conic be given—the ellipse may be the easiest case—and fix a particular diameter $ED$. Let $A$ be a point on diameter $ED$ extended. A special line is paired with $A$, so special that it is called the *polar* of $A$ (with respect to the conic), with $A$ called the *pole* of the line. We trace the polar of $A$ by drawing the two tangents from $A$ to the conic, meeting the conic at $K$ and $L$. Line $KL$ is the polar of $A$. Further, $KL$ is an ordinate with respect to diameter $ED$. Further still, if we draw any secant on $A$, meeting the conic at $H$ and $N$, then $H(AG, HN)$ where $G$ lies on $KL$. And still more, tangents to the conic at $H$ and $N$ meet on the polar $KL$.

A summary is included in this broader theorem; for proofs we will later follow the work of Philippe de la Hire.

**Fig. 4.3**  *A* is the Pole of line
*KL*.
$H(AM, ED)$,  $H(AG, HN)$,
and $H(BG, KL)$

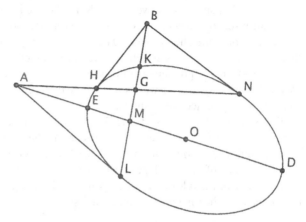

**Theorem 4.5**

(1) *When a tangent to a conic at a point L meets a diameter ED extended at point A, then H(AM, ED), where the ordinate KL, corresponding to diameter ED, meets the diameter at M.*

(2) *Let the tangents from outside point A meet the conic at L and K, and let a secant on A meet the conic at H and N, and meet KL at G. Then H(AG, HN), and the tangents at H and N meet on line KL.*

(3) *The ordinates for a diameter ED are parallel to the tangents at vertices E and D (proved above as Theorem 4.2, which means tangents at vertices E and D are parallel to each other.*

The *pole-polar* relation can be extended to the case when the pole lies inside a conic. Both cases, the pole inside and the pole outside the conic, are summarized in the following definition.

**Definitions X** Given a conic section *c*, possibly a circle, and an outside point *A*, the *polar* of *A* (with respect to the conic) is the line *KL* where tangents to *c* from *A* meet the conic at *K* and *L*. *A*, in turn, is the *pole* of line *KL*.

When a point *G* is inside a conic, we say line *p*, outside the circle, is the *polar* of *G* if for every point *X* of *p*, the polar of *X* lies on *G*. In this case we also say *G* is the *pole* of *p*.

Since tangents at the endpoints of a diameter are parallel, the *polar* of the center of a conic is the line at infinity.

## 4.5  Late Antiquity and Steps Toward Projective Geometry

This introduction to the *Conics* of Apollonius covers only part of the first of eight books. Further propositions will be covered in later chapters, in the methods of later mathematicians.

Why do we pay attention to the *Conics* of Apollonius in discussing projective geometry? To apply projective methods to the study of conic sections means, above all else, to consider the conic section as the projection of the base circle of the cone onto the slicing plane. The projective methods carry to conic sections those properties of circles which are preserved under projection. The methods of Apollonius were not projective in this sense. However, the definitions and theorems of the *Conics* of Apollonius were still the core definitions and theorems when conics were studied by projective methods.

There were advances by later Greek authors related to conic sections. Of particular importance is Menelaus's Theorem, the first proposition of Book 3 of Menelaus's *Sphaerica*, of about 100 AD, although probably known earlier. It was a theorem of spherical geometry, whose immediate application was in the trigonometry of triangles on a sphere whose sides are great circles. (Trigonometry developed after the time of Apollonius, as astronomy became highly mathematical.)

Here is the planar version of Menelaus's Theorem. The points are named as they would be by Girard Desargues, who made crucial use of that theorem.

**Theorem 4.6**  *If collinear points D, H, G lie, respectively, on sides (extended)* $4h$, *$hK$, and $K4$ of triangle $Kh4$, then*

$$\frac{Dh}{D4} = \frac{Hh}{HK} \cdot \frac{GK}{G4}.$$

The proof is by triangle similarity in Fig. 4.4, where *KF* is drawn parallel to line *HDG*.

Later, we have the *Mathematical Collection*, by Pappus, from the early fourth century AD, a survey in eight books of mathematical topics. It reports on some noted works which have since disappeared, including a couple by Euclid. In Pappus's Book 8 is found the construction of a conic through five given points. We also have in Book 7 Prop. 238 the focus-directrix definitions of the conic sections: given a line (directrix) and a point *C*, and taking the points *D* produced so the distance *DC* is to the distance of *D* from the directrix in a given ratio, then *D* is on a conic. The conic is an ellipse, parabola, or hyperbola as that given ratio is less than, equal to, or exceeding 1.

Further, Pappus proved, in his Book 7, the Pappus Hexagon Theorem, one of the foundational propositions of projective geometry. This theorem appears with his lemmas to Euclid's lost work, the *Porisms*. Pappus's Hexagon Theorem states that if the vertices of a hexagon lie alternately on two lines, then the three pairs of opposite sides meet in collinear points. Figure 4.6 Left illustrates the theorem, and Exercise 2 outlines a proof—not that of Pappus. We will prove the theorem again in the chapter on Jacob Steiner.

Menelaus's Theorem and Pappus's Hexagon Theorem involve collinear points. Theorems involving collinear points or concurrent lines are almost entirely absent from Euclid's *Elements*, and rare in the *Conics* of Apollonius. Projective geometry, on the other hand, is quite concerned with collinear points and concurrent lines.

**Fig. 4.4**  Menelaus's Theorem

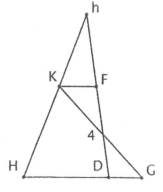

We have already seen several such theorems, including Desargues' Theorem and its converse. Pappus, especially, states and proves theorems of a projective nature.

Here is one of Euclid's *Hyptios Porisms*, reported by Pappus, showing that Euclid's work did include  consideration of points collinear and lines concurrent. The wording is modernized, from [9, p. 556].

> Let three variable straight lines $l_1, l_2, l_3$ meet straight line $l_4$ in, respectively, unvarying points $P_1, P_2, P_3$, while the intersection of $l_2$ and $l_3$ lies on a given straight line $m_2$ and the intersection of $l_1$ and $l_3$ lies on a given straight line $m_1$. Then it is possible to construct a straight line $m_3$ on which the intersection of $l_1$ and $l_2$ lies.

In Fig. 4.5, $P_2$, $A'$ and $J$ are collinear, and $m_3$ will be the line on $I$, $J$, and $L$. We will see a proof after we examine the work of Jacob Steiner.

The early modern European mathematicians who developed projective methods in their geometry had studied Pappus. Lemmas that can be considered preliminary to Girard Desargues' concept of involution are found in Book 7 of Pappus's *Mathematical Collection*, and, as noted, Girard Desargues, made great use of the Theorem of Menelaus. After the flurry of projective work of seventeenth century geometers Desargues, Blaise Pascal, and Philippe de la Hire, interest in projective methods subsided for a century. But when it came back, beginning at the end of the eighteenth century, it came with a flurry of theorems about collinear points and concurrent lines.

**Fig. 4.5** A Hyptios Porism, of Euclid, as reported in Pappus's *Mathematical Collection*, Book 7

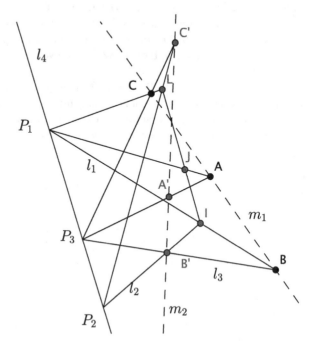

## 4.6 Notes and Exercises

**Note 1** The *Conics* of Apollonius is a demanding book. Over a century ago, Sir Thomas Heath wrote a compacted edition in English [5], in the symbols of modern algebra. We are warned on the title page that the work is "edited in modern notation." And he devoted 70 pages of his general history of Greek mathematics [6] to Apollonius. This may still be the best place to start. The first three books of the *Conics* were translated, respecting the presentation of Apollonius and without compacting, by Catesby Taliaferro, reissued in 1998 [1]. At the time of Heath, there was controversy among scholars as to whether Greek mathematicians had put geometric clothing on discoveries made by some "geometric algebra." There is now consensus that their work is truly geometric, and this case is strongly made in [4], where there is also a nice argument that the conics were, for Apollonius, truly slices of a conic surface in space, and not, again, objects defined by equations.

**Note 2** I call Eqs.(4.1), (4.2), and (4.3) *abscissa-ordinate* equations. Among ancient Greeks, they were *symptoma*. They have much in common with modern coordinate equations, but they are significantly different. Greeks did not make the plane into a coordinate plane. Their *symptoma* related an area involving length along the diameter of a conic to the square on the corresponding ordinate. They would not, for example, give equations of two conic sections with respect to the same coordinate axes.

**Exercise 1** The parabola is formed when the slicing plane makes $EG$ parallel to $AC$, in Fig.4.1 Left. Show that by adjusting the argument from the case of the ellipse or hyperbola, we reach the equation:

$$ML^2 = EM \cdot Constant1 = EM \cdot 4p, \quad for \ p = \frac{BG \cdot}{4EG} \cdot$$

**Exercise 2** Below is *Pappus's Hexagon Theorem*, with an outline of a proof as found in [2]. The reader is asked to complete the proof.

**Theorem 4.7 (Pappus Hexagon Theorem)** *Let the opposite sides of a hexagon $ABCDEF$ lie alternately on two lines, $ACE$ and $BDF$. Then the opposite sides meet in collinear points. (See Fig. 4.6 Left.)*

**Proof** Let $G = AC \cap BD$. Let opposite sides $AB$ and $DE$ meet in $X$, opposite sides $BC$ and $EF$ meet in $Y$, and opposite sides $CD$ and $AF$ meet in $Z$. We apply a central collineation mapping line $XY$ to the line at infinity. This gives the diagram of Fig. 4.6 Right. So we have $AB \parallel DE$ and $BC \parallel EF$. Finish the proof, using the Side Splitter Theorem. □

This theorem is proven in Book 7 of Pappus's *Mathematical Collection*. We'll outline Pappus's argument using the concept of a cross-ratio in Chap. 10. In Coxeter's [3, pp.231–238], Pappus's Hexagon Theorem is accepted as an axiom, and then Desargues' Theorem is proven. We need assume neither Desargues' Theorem nor Pappus's Theorem as an axiom because our projective plane is in 3-

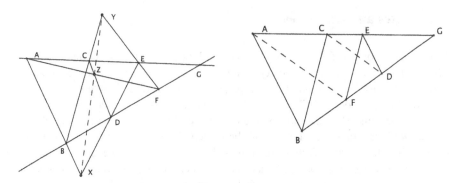

**Fig. 4.6** Left: Exercise 2 Pappus's Theorem. Right: Exercise 2 figure transformed

dimensional space, in which we have plane-to-plane projection. The proof here uses the Side Splitter Theorem of Euclidean geometry. An alternate proof, not dependent on Euclidean geometry, is also given in the chapter on Steiner (Chap. 10).

## 4.7   Some Solutions

**Exercise 1**   Solution: Since line $EMG$ is parallel to $AC$,

$$\frac{PM}{EM} = \frac{BG}{EG.}$$

See Fig. 4.1 Left. In the circle on $P$, $L$, and $R$,

$$ML^2 = PM \cdot MR = \frac{BG \cdot EM}{EG} \cdot MR = EM \cdot 4p, \text{ for } p = \frac{BG \cdot}{4EG}.$$

**Exercise 2**   Completion of the proof of Theorem 4.7.

Because $AB \parallel DE$, then by the Side Splitter Theorem,

$$\frac{AG}{BG} = \frac{AE}{BD} = \frac{GE}{GD.}$$

And since $BC \parallel EF$, then

$$\frac{CG}{BG} = \frac{CE}{BF} = \frac{GE}{GF.}$$

Thus

$$\frac{GC}{GD} = \frac{BG \cdot CE}{BF} \cdot \frac{AE}{BD \cdot GE} = \frac{BG \cdot AE}{BD} \frac{CE}{BF \cdot GE} = \frac{GA}{GF}.$$

We conclude that $CD \parallel AF$, so $X$, $Y$, and $Z$ are collinear on the line at infinity. Since a central collineation preserves collinearity, then $X$, $Y$, and $Z$ are collinear in the original given hexagon.□

# References

1. Apollonius of Perga, *Conics*, transl. R. C. Taliaferro, Santa Fe, NM: Green Lion Press, 1998.
2. Richard Courant and Herbert Robbins, *What is Mathematics?*, New York and London: Oxford University Press, 1941; reissued, revised by Ian Stewart 1996.
3. H. S. M. Coxeter, *Introduction to Geometry*, New York: Wiley, 1969.
4. Michael Fried and Sabetai Unguru, *Apollonius of Perga's Conics: Text, Context, Subtext*, Leiden: Brill, 2001.
5. Sir Thomas Heath, *Apollonius of Perga Treatise on Conic Sections*, Cambridge University Press, 1896; reissued by Palala Press, 2015.
6. Sir Thomas Heath, *A History of Greek Mathematics*, Oxford: Clarendon Press, 1921; reissued by Dover, New York, 1981.
7. Victor Katz, *A History of Mathematics: An Introduction* 3rd edition, Boston: Addison-Wesley, 2009.
8. Claude Mydorge, *Prodromi catoptricorum et dioptricorum: sive conicorum operis ad abdita radii reflexi et refracti mysteria praevii et facem praeferetis*, Books 1 and 2, 1631, Books 3 and 4, 1639.
9. Pappus of Alexandria: Book 7 of the *Collection*, edited by Alexander Jones. New York: Springer-Verlag, 1986.
10. Ivor Thomas, *Selections Illustrating the History of Greek Mathematics*, Vol 1, Cambridge, MA: Harvard University Press, 1939.

# Chapter 5
# Conic Sections in Early Modern Europe.
# First Part: Philippe de la Hire on Circles

The early modern study of conic sections began with the publication of edited translations, into Latin, of the first four books of the *Conics* of Apollonius in the sixteenth century, especially the translation by Commandinus, of Urbino, in 1566. Books 5, 6, and 7 were later found in Arabic manuscripts. Book 8 seems lost, although that loss has prompted several mathematicians, over the centuries, to propose their own restorations.

To bring an ancient text to a Latin edition was not simply a work of translation. The translator/editor worked from several manuscripts, usually incomplete, often copies of copies made by persons who did not understand the mathematics. The translator/editor needed to be a mathematician, writing coherent mathematics and supplying proofs where they were missing or defective. In that same century, some shorter original works were published, notably Johannes Werner's [17] and Francesco Maurolico's [11]. In the seventeenth century, we have comprehensive works, all based on the *Conics* of Apollonius, all aiming to simplify the demonstrations of Apollonius. Most prominent are Claude Mydorge's [12], of 1631, followed by [13] of Gregory St. Vincent, and John Wallis's [16]. In 1639, a short work by Girard Desargues was published in 50 copies, with the title *Brouillon project d'une atteinte aux événements des rencontres d'un cône avec un plan*, henceforth referred to as *Brouillon project* [6]. "Brouillon" means a "sketch," which is good to keep in mind. The title could be translated, "A draft study of the sections of a cone by a plane." It was thoroughly projective, assuming a plane with a line at infinity, where parallel lines meet at a point on this line at infinity, and carrying, by projection, properties of the base circle to the conic section.

Desargues was a highly original mathematician. So original that all copies of the *Brouillon projet* seem to have disappeared within decades after its publication, a mark of its difficulty and unusual character. Desargues and his ideas did have some influence in the circle of scholars who met regularly with Marin Mersenne, in Paris. This informal academy included Mydorge, Desargues, Gilles de Roberval, Étienne Pascal and his son Blaise. By correspondence, Mersenne had ties to Descartes,

© Springer Nature Switzerland AG 2020                                   59
C. Baltus, *Collineations and Conic Sections*,
https://doi.org/10.1007/978-3-030-46287-1_5

Fermat, and the other major scientists of Europe. The young Pascal, in particular, was influenced by Desargues; he wrote in *Essay pour les coniques* of 1640, "I have tried as far as I could to imitate [Desargues'] method of approaching this material."[7, p. 183]. The *Brouillon project* was transcribed in 1679 by Philippe de la Hire. That transcription, in turn, disappeared until 1845. The existence of Desargues' work was never in question, since it was discussed in various letters which survived, and was at the center of a vitriolic dispute in the 1640s. See [15].

## 5.1  Philippe de la Hire

Our examination of seventeenth century work will center on publications of Philippe de la Hire (1640–1718), especially his *Nouvelle Méthode en Géométrie pour les Sections des Superficies coniques et Cylindriques* of 1673. Although La Hire lived a generation later than Desargues, there was a connection to Mersenne's circle, in that Philippe's father, the painter Laurent de la Hyre (1606–1656), was a good friend of Desargues, and Laurent's attention to perspective is attributed to this friendship. Further, in the 1670s, Philippe worked with Abraham Bosse (1604?–1676), a collaborator of Desargues.

Philippe worked in the tradition of Mydorge in simplifying the *Conics* of Apollonius, and at least some influence by Desargues' work is inevitable. (Reports that Philippe de la Hire was a "disciple" of Desargues are exaggerated.)

Philippe de la Hire trained to be a painter, the profession of his father, but moved to mathematics. Details about this change are limited. Bernard de Fontenelle, in his short *Éloge* of 1718, reported that the teenage Philippe was interested in geometric aspects of painting, including perspective. In Italy, from 1660 until 1664, he developed his art, but also found a love for Greek geometry, especially the *Conics* of Apollonius. After 1664, La Hire apparently continued his geometric studies. He must also have continued in painting, for he was admitted to the painters' guild, *l'Académie de Saint-Luc*, in 1670. He tells us at the beginning of his book of 1673 [9] that he had applied himself to the study of geometry for "several years," but we know nothing specific until 1672, the year of his first publication in mathematics [8].

That short work of 1672 comprises seven propositions with corollaries. It was brought out by Abraham Bosse. The problem addressed is that of constructing an *arc rampant*, and that problem was, it seems, involved in a controversy between Bosse and (Nicolas)-François Blondel (1618–1686). The problem is to construct a connected piece of a conic where two endpoints, and tangents at those endpoints, are given, together with one additional tangent line. Blondel, familiar with the *Conics* of Apollonius, solved the *arc rampant* problem in a 1673 work on architecture [3]. (See [1].)

Bosse, a master of etching, had worked with Girard Desargues in bringing a mathematical foundation to the professions. Working with rules found by Desargues, he brought out in 1643 the first part of a treatise on stone cutting [4].

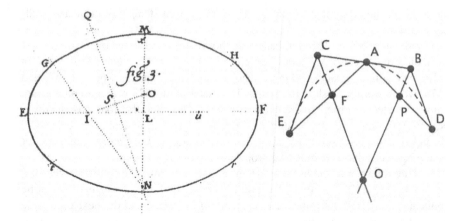

**Fig. 5.1** Left: from Bosse, *Traité des practiques géométrales* . . ., 1665. Right: based on *Fig* 1, La Hire 1672

When he turned to producing a second volume on stone cutting, long after the death of Desargues, Bosse realized the need for a mathematical collaborator on construction of the *arc rampant*. Bosse turned to Philippe, the son of his friend, Laurent. The result was [8]. In that 1672 work, La Hire employed classical methods. This is puzzling since he was already working on the 1673 book in which much simpler methods, of a projective nature, were developed. See Fig. 5.1 Right, based on La Hire's [8]. In particular, an arc of a conic is sought with given endpoints *D* and *E*, with given tangents to the conic at *D* and *E*, and with additional tangent line, *BC*, given. We'll suppose that *BC* meets *DE* at a point *K*, not in the diagram. The crucial step is to find the point of tangency, *A*, on *BC*. Both La Hire and Blondel concluded that *A* is found on *ED* so *K* and *A* are harmonic conjugates to points *B* and *C*. (The term "harmonic" was introduced in the works of 1673 by both La Hire and Blondel.)

The dispute between Blondel and Bosse seems to have centered on Bosse's acceptance of approximate methods. For example, he suggested an approximation of an ellipse by a pair of congruent small circular arcs joined smoothly by arcs from a larger circle, as in Fig. 5.1 Left. La Hire gave no indication that he had joined this dispute. In [9], although he had presented in Lemma 17, with respect to a circle, what was needed to find the *arc rampant*, he only referred to the *arc rampant* on the last page.

In addition to his *Nouvelle Méthode* of 1673, La Hire issued two other comprehensive works on conic sections. *Nouveaux Elemens des Sections Coniques* was published in 1679. The three conic sections were introduced separately, based on the foci definitions that will be familiar to modern calculus students. For example, the ellipse on major axis *IT*, with foci *F* and *H* on that axis, is defined as the path of those points *X* so that the sum $XF + XH$ is *IT*. Then the abscissa-ordinate equations were derived, although using the algebra of ratios and computations by areas, in flavor similar to that of Apollonius, even if notably clear and well-organized.

Finally, in 1685, La Hire published his most complete, best known, and longest work on conic sections, *Sectiones Conicae en novem libros distibutae* [10]. It was in Latin, indicating a broader and more scholarly audience than his works of 1673 and 1679, which were published in French. The material of 1673 was presented again, with some cleaner proofs and additional results. This time the goal was clearly to present Apollonius' results, simplified, reorganized, and, in several cases, expanded. He ended with a brief summary of the first seven books of the *Conics*, indicating for each proposition its corresponding proposition in the 1685 work.

La Hire returned to mathematics from time to time over the following decades, but his career shifted to that of a public servant in a broad range of scientific applications. In 1678 he joined the Academy of Sciences, and the next year made the first of several expeditions in the mapping of France, a project of Minister Colbert. He directed water projects, including one to bring water from the vicinity of Chartres to Versailles. He published tables of sun and moon positions, based on observations at the Paris Observatory, where he lived from 1682. His interests extended to architecture; he was appointed to the Royal Academy of Architecture in 1687, taking the seat of Blondel, who had died the year before, and some years later issued an influential manuscript on stone cutting for vaults [2, pp. 321–325].

## 5.2   On Circles: La Hire's First 17 Lemmas of 1673

Harmonic division, which we introduced in Chap. 4, was so central to La Hire that he opened his [9] of 1673 by defining the expression:

> I call the straight line $AD$ cut in 3 parts harmonically when the rectangle contained by all $AD$ and the middle part $BC$ is equal to the rectangle contained by the two extreme parts $AB, CD$.

Philippe de la Hire's 1673 introduction to the conic sections began with 17 lemmas that develop the pole and polar concept in the case of a circle, based on the harmonic division of a secant to a circle (Fig. 5.2).

La Hire's Lemmas 2 through 6 show that projection of a harmonic set on one line produces a harmonic set on a second line. Parallels are treated separately. The only suggestion that parallel lines are in some sense concurrent is in the diagrams: in La Hire's *Fig.* 6 of [9], four parallel lines on $C$, $D$, $E$ and $F$ are labeled $AC$, $AD$, $AE$, and $AF$.

La Hire's Lemma 1 is the construction, seen in La Hire's Fig. 5.1, of point $B$, between $A$ and $C$ when $A$, $C$, and $D$ are given in that order, so $H(AC, BD)$. A line $DG$ is drawn on $D$, then $FCE$ is drawn on $C$ parallel to $AG$, with midpoint $C$. Then join $F$ and $G$. $FG$ meets $AD$ at $B$, the point sought.

*Proof* By similar triangles,

**Fig. 5.2**  La Hire $Fig.$ 1, 1673

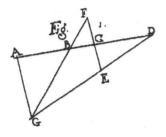

$$\frac{AD}{CD} = \frac{AG}{CE} = \frac{AG}{FC} = \frac{AB}{BC}.$$

Of particular importance is the case, Lemma 3, where segment $AD$, with $H(AC, BD)$, is projected from a point $E$, with $H(AC, BD)$, onto a line $AG$ that is parallel to $ED$. We would now say that in this case $D$ is mapped to infinity. This occurs exactly when $G$, the image of $B$, is the midpoint of $AH$, where $H$ is the image of $C$. We give a partial proof, based on LaHire's proof.                                    □

**Theorem 5.1 (La Hire's Lemma 3, 1673)** *Suppose $H(AC, BD)$ on line $AD$ and $AD$ is projected from a point $E$ onto line $AH$ that is parallel to $ED$, $B$ to $G$ and $C$ to $H$. Then $G$ is the midpoint of $AH$.*

**Proof**  See Fig. 5.3. By similar triangles,

$$\frac{AG}{DE} = \frac{AB}{BD} \text{ and } \frac{AH}{DE} = \frac{AC}{CD}.$$

Solving for $DE$ in each equation gives

$$\frac{BD \cdot AG}{AB} = \frac{CD \cdot AH}{AC}. \text{ So } \frac{AH}{AG} = \frac{BD \cdot AC}{AB \cdot CD}.$$

The numerator is
$(CB + CD)(AB + CB) =$
$AB \cdot CD + CB(AB + BC + CD) =$
   $AB \cdot CD + CB \cdot AD$, so division by
$AB \cdot CD = AD \cdot CB$ shows that $AH/AG = 2$.                                    □

We can, alternately, arrive at the Lemma 3 by treating $H(AB, MC)$ as the equality of ratios

$$\frac{AC}{MB} = \frac{AM}{BC}.$$

Fixing $A$ and $B$, as $C$ approaches infinity, $M$ must approach the midpoint of $AB$. Here is the result restated as a theorem:

**Fig. 5.3**  based on La Hire's
Lemma 3, 1673

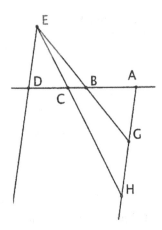

**Theorem 5.2 (Infinity in a Harmonic Set)** *Given finite points A and B, the midpoint of segment AB is the harmonic conjugate of the point at infinity on line AB.*

La Hire's proofs of the invariance of harmonic sets under projection, in Lemmas 2–6, are, as in the case above, exercises in similar triangles and properties of proportion. I substitute a simpler proof used by Lazare Carnot in 1806 [5, p. 93]; Carnot's method can also show the invariance of the cross-ratio and other expressions, as we will see later.

**Theorem 5.3 (La Hire's Lemma 5, 1673: Invariance of the Harmonic Relation)** *Let $A, B, C, D$ be a harmonic set with $B$ between $A$ and $C$, i.e., $H(AC, BD)$, projected from a point $E$ to line $FH$: $A$ to $F$, $B$ to $G$, $C$ to $H$ and $D$ to $I$. Then $H(FH, GI)$.*

***Proof***  See Fig. 5.4. By the Law of Sines: in $\triangle EBA$,

$$\frac{sin(\angle BEA)}{sin(\angle EAB)} = \frac{BA}{EB}.$$

We'll show

$$\frac{AD \cdot BC}{AB \cdot CD} = \frac{FI \cdot GH}{FG \cdot HI};$$

this would mean that the harmonic property is preserved.
We have

$$\frac{AD}{AB} = \frac{ED}{EB}\frac{sin(\angle AED)}{sin(\angle EAD)}\frac{sin(\angle EAB)}{sin(\angle AEB)} = \frac{ED}{EB}\frac{sin(\angle AED)}{sin(\angle AEB)}$$

since $\angle EAD = \angle EAB$. Likewise,

**Fig. 5.4** Based on La Hire's
Lemma 5, 1673

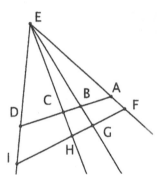

$$\frac{BC}{CD} = \frac{EB}{ED}\frac{sin(\angle BEC)}{sin(\angle CED)}$$

since $\angle ECD = \angle ECB$ (but in the opposite sense—which was important to Carnot).
   Since the ratio depends only on the angles formed by the four lines at point $E$,

the same ratio equals $\dfrac{FI \cdot GH}{FG \cdot HI}$.                                            □

   Poncelet, in Art. 20 of his 1822 *Traité*, provided a more general rule for invariance under projection. We state the claim as a theorem.

**Theorem 5.4 (Carnot's Principle)** *Suppose we have a ratio involving collinear points that can be expressed as ratios of the sines of angles from an outside point $E$ drawn to those collinear points. As with Carnot's argument, those ratios will be invariant under projection from $E$.*

   Lemma 7 concerns two harmonic sets which share a point.

**Theorem 5.5 (La Hire's Lemma 7, 1673)** *Suppose $H(BG, FH)$ and $H(BD, CE)$, on two lines which meet at $B$. Then the lines $GD$, $FC$, and $HE$, on the three pairs of corresponding points, are concurrent (or parallel—a case treated separately by La Hire).*

***Proof in the Non-parallel Case*** See Fig. 5.5, La Hire's *Fig.* 8 of 1673. We let lines $EH$ and $FC$ meet at $A$, and we suppose $GD$ does not lie on $A$, but rather $AD$ meets line $BH$ at $L$. So $H(BL, FH)$ since, by La Hire's Lemma 5, a harmonic set is projected to a harmonic set. So we have $H(BG, FH)$ and $H(BL, FH)$. But this is only possible if $G = L$.                                            □

   In La Hire's Lemmas 8 and 9, we consider a point $A$ outside a given circle, where tangents from $A$ meet the circle in points $F$ and $G$. See La Hire's *Fig.* 12 and 13 of 1673. In our Fig. 5.6 La Hire's *Fig.* 12, $A$ is on diameter $BD$. Then $H(AC, BE)$, shown by similar triangles $ALB$ and $AHE$, the tangent property: $LB = LF$ and $HF = HE$, and a parallel projection from $AH$ to $AE$. This is Lemma 8.
   In Lemma 9, the result of Lemma 8 is extended to any line on $A$ meeting the circle in $O$ and $L$ and meeting the polar in $I$. ( La Hire's *Fig.* 13.) La Hire provided

**Fig. 5.5**  La Hire's *Fig.* 8,
for Lemma 7, 1673

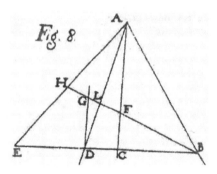

a three-dimensional proof. The sphere for which the given circle is a great circle is
cut perpendicular to the plane $AGE$, giving a circle for which $OL$ is a diameter and
$DH$ the polar of $A$. So by Lemma 8, $H(AI, OL)$.

Here is a two-dimensional proof. See Fig. 5.6 Bottom Right. Draw the circle with
diameter $OL$, and from $A$ draw tangents $AK$ and $AJ$ to this circle. Since $O$ and $L$
are common to the two circles, the power-of-point $A$ to those circles is the same,
$AO \cdot AL$. So tangent segments are equal: $AF = AG = AJ = AK$. So $F, G, J, K$
all lie on a circle with center $A$. Now $I$ lies on $JK$ since $I$ is the intersection of
the common secants of the three circles in the figure. So by La Hire's Lemma 8,
$H(AI, OL)$.

In summary,

**Theorem 5.6 (La Hire's Lemmas 8 and 9, 1673)** *Let $A$ be a point outside a circle,
where $GF$ is the polar of $A$, i.e., tangents from $A$ meet the circle at $G$ and $F$. Then
if a line on $A$ meets the circle in points $B$ and $E$ and meets $GF$ in point $C$, then
$H(AC, BE)$. Further, since three points of a harmonic set determine the fourth
point, a point $C$ inside the circle is on the polar of $A$ exactly when line $AC$ meets
the circle in $B$ and $E$ where $H(AC, BE)$.*

In Lemma 10, the inscribed quadrilateral $BOLE$ has sides $BE$ and $OL$
meeting at $A$. We have harmonic sets which share point $A$, so by Lemma 7, their
corresponding points are joined by concurrent lines—here concurrent at $D$, on the
*polar* of $A$. Note that $D$ may lie inside the circle. See Fig. 5.7, based on La Hire's
*Fig.* 14 of 1673.

La Hire continued Lemma 10, noting that when point $O$ coalesces with $B$, and,
likewise, point $L$ coalesces with $E$, then $BO$ and $EL$ become tangents to the circle
at $B$ and at $E$, still meeting on the polar of $A$.

In summary:

**Theorem 5.7 (La Hire's Lemma 10)** *Let $A$ be outside a circle, and a line on $A$
meet the circle in points $B$ and $E$. Then tangents at $B$ and $E$ meet on the polar of
$A$. And if a second line on $A$ meets the circle at points $O$ and $L$, then the points of
intersection $BO \cap EL$ and $BL \cap EO$ lie on the polar of $A$. This means that when
point $D$, outside the given circle, lies on the polar of $A$, then the polar of $D$ lies
on $A$*

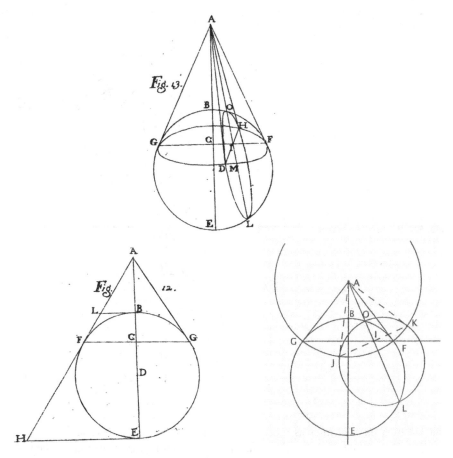

**Fig. 5.6**  Top: La Hire's Fig 13. for Lemma 9, 1673. Bottom Left: La Hire's Fig. 12 for Lemma 8, 1673. Bottom Right: New proof of La Hire's Lemma 9

The case when a point $C$ is inside the given circle, but not its center, is covered in La Hire's Lemmas 13 and 14. See La Hire's *Fig. 19*, our Fig. 5.8. A proof based on central collineations appears in Chap. 7.

**Theorem 5.8 (La Hire's Lemmas 13 and 14, 1673)**  *Let $C$ be inside a given circle, not its center. Let $BD$ be the diameter on $C$, extended to $A$ so $H(AC, BD)$. On $A$ draw line $AL$ perpendicular to $BD$. Draw line $LC$, meeting the circle in $F$ and $E$. Then $H(LC, FE)$. This means line $AL$ is the polar of $C$, and it is unique since no other point $L$ outside the circle satisfies $H(CL, FE)$. Conversely, $C$ is the pole of line $AL$.*

La Hire did not use any terminology for the *pole-polar* relationship, suggesting he did not recognize the power of the concept. He saw no need to consider the case of a point lying on the circle: its polar would be the tangent at that point. That he

**Fig. 5.7** Based on La Hire's
Lemma 10, 1673

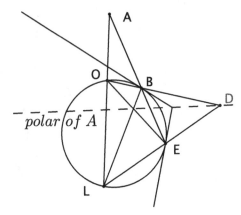

**Fig. 5.8** Based on La Hire's
*Fig.* 19, 1673

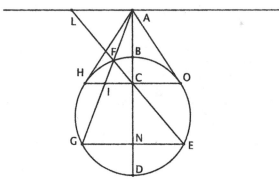

did not treat parallel lines as meeting on a line at infinity, an idea that Desargues
exploited, complicated the matter with a proliferation of cases. Nevertheless he has
shown the heart of the matter:

**Theorem 5.9 (Pole-Polar Duality)** *Any point X lies on line a, the* polar *of a point
A, exactly when x, the* polar *of X, lies on A.*

## 5.3   Notes and Exercises

**Exercise 1** Give details of the proof of La Hire's Lemma 8 of 1673, the case,
pictured in Fig. 5.6 Bottom Left, that $H(AC, BE)$ when $BE$ is a diameter of the
circle. Note: It follows from Euclid's Book 3 Prop. 36 that the tangent segments
from an outside point to a circle are of equal length.

**Exercise 2** Now that we know that $H(AB, C\infty)$ exactly when $C$ is the midpoint
of segment $AB$ and that a central collineation maps a harmonic set to a harmonic
set, we justify by a central collineation Jacob Steiner's method for constructing by

**Fig. 5.9** Steiner
Construction, 1833,
Exercise 2

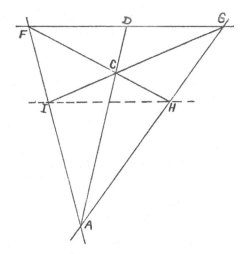

straightedge alone a parallel to a given line on a given outside point, provided we already have three equally spaced points on the given line [14, Chapter 1 Art. 8].

Here is the construction when we have points $F$ and $G$, with midpoint $D$, on the given line, with $H$ the given outside point. See Fig. 5.9. We draw lines $FH$ and $GH$, and on $GH$, beyond $H$, pick a point $A$ and join it to $D$ and to $F$. Let $C = FH \cap AD$. Let $GC$ meet $AF$ in $I$. Then $HI \parallel FG$. Why?

Here is the outline of a proof. We apply the central collineation with center $C$, where the axis is the line on $A$ parallel to $FG$, and $F$ is mapped to $H$.

a. Why is the image of line $FG$ a parallel line on $H$?
b. Why is $D'$, the image of $D$, the midpoint of $G'H$?
c. Why is $G'$ on $FA$? Note that any parallel to $FG$ is divided equally by lines $AF$, $AD$, and $AG$.
d. Why is $G'$ equal to $I$? This means $HI \parallel FG$.

# References

1. Christopher Baltus, The arc rampant in 1673: Abraham Bosse, François Blondel, Philippe de la Hire, and conic sections, *Research in the History and Philosophy of Mathematics, The CSHPM 2015 Annual Meeting*, 2016, Cham, Switzerland: Birkhäuser/Springer International, 15–24.
2. Edoardo Benevenuto, *Introduction to the History of Structural Mechanics: Part II: Vaulted Structures and Elastic Systems*, New York: Springer, 2012.
3. (Nicolas)-François Blondel, *Résolution des quatres principaux problèmes d'architecture*, Paris: Imprimerie Royale, 1673.
4. Abraham Bosse, *La pratique du trait à preuves de M. Desargues, Lyonnois, pour la coupe des pierres en l'architecture*, Paris: Pierre Des-Hayes, 1643.
5. Lazare Carnot, *Essai sur la Théorie des Transversales*, Paris: Courcier, 1806.

6. Girard Desargues, *Brouillon project d'une atteinte aux événements des rencontres d'un cône avec un plan*, in [7] (trans. Field) and original http://gallica.bnf.fr/ark:/12148/bpt6k105071b/f1.image Paris, 1639.

7. J. V. Field and J. J. Gray, *The Geometrical Work of Girard Desargues*, New York: Springer, 1987.

8. Philippe de La Hire, *Observations de Ph. de la Hire sur les points d'attouchement de trois Lignes droits qui touchent la Section d'un Cone sur quelques-uns des Diametres, et sur le centre de la mesme Section*, Paris: A. Bosse, 1672.

9. Philippe de La Hire, *Nouvelle Méthode en Géométrie pour les Sections des Superficies coniques et Cylindriques*, Paris, 1673.

10. Philippe de La Hire, *Sectiones Conicae en novem libros distributae*, Paris 1685; French translation by Jean Peyroux, *Grand Livre des Sections Coniques*, Paris: Blanchard, 1995.

11. Francesco Maurolico, *Opuscula mathematica: nunc primum in lucem aedita, cum rerum omnium notatu dignarum...* Venis: Franciscum Farnciscium Senensem, 1575.

12. Claude Mydorge, *Prodromi catoptricorum et dioptricorum: sive conicorum operis ad abdita radii reflexi et refracti mysteria praevii et facem praeferetis*, Books 1 and 2, 1631, Books 3 and 4, 1639.

13. P. Gregorii a Sto. Vicentio, *Opus geometricum quadraturae circuli et sectionum coni*, Antwerp: Ioannem and Iacobum Meursios, 1647.

14. Jacob Steiner, *Die geometrischen Konstructionen, ausgeführt mittelst der geraden Linie und Eines festen Kreises, ...*, Berlin: Dümmler, 1833, translated and edited by M. E. Stark and R. C. Archibald, as *Jacob Steiner's Geometrical Constructions with a Ruler*, New York: Scripta Mathematica, 1950.

15. René Taton, *L'oeuvre mathématique de G. Desargues*, Paris: Presses Universitaire de France, 1951.

16. John Wallis, *De sectionibus conicis, Nova methodo expositis, tractatus*, 1655, in Wallis, *Opera Mathematica*, Oxford, 1695, reissued Hildesheim: Georg Olms, 1972.

17. Johannes Werner, *Super vigintiduobus Elementis Conicis*, Vienna: Lucae Alentsee Bibliopolae, 1522.

# Chapter 6
# Conic Sections in Early Modern Europe. Second Part: Philippe de la Hire on Conics

After the preliminary work on circles, La Hire moved to the conic sections. He defined them as had Apollonius: the section of a conic surface by a plane is a conic section. But he immediately moved beyond Apollonius, as had Desargues, by claiming that properties established for the base circle apply to the conic section. La Hire wrote in [5, p 15] of 1673, following the first seventeen lemmas, "All that follows is a simple application of these lemmas ... in all the conic and cylindrical sections." The harmonic relation is preserved in a projection, and the projection he has in mind is of a base circle of a cone, in a base plane, and lines in that base plane associated with that circle, onto the plane that cuts the cone in the conic section. La Hire describes the projection of a line $eg$ in the base plane: "cut at points $e$, $f$, $p$, $g$ in three harmonic parts. ... the lines drawn joining these points of division to the vertex $A$ [of the cone] will meet line $EG$ on the slicing plane in points $E$, $F$, $P$, $G$ and by Lemma 5 or 6 it will be cut in these points $E$, $F$, $P$, $G$ harmonically in 3 parts. ... "

Therefore our Theorems 5.6 through 5.9, about circles, are proven for a conic.

Although not given as the definition, Desargues and La Hire conceived of a conic section as the projection of a circle onto a plane. We state this in terms of a central collineation, with a few related definitions.

**Definitions XI** A *conic section* is the image of a circle under a central collineation.

Given a conic surface with vertex $A$ and a slicing plane, the plane on $A$ that is parallel to the slicing plane is the *vertical plane* or *planum verticale*. (This concept was used by Desargues and by La Hire in 1673 [5], but the name is from La Hire's [7] of 1685.) Plane $ABC$ in Fig. 6.6 is the vertical plane. The vertical plane meets the base plane in the *vanishing line* of the projection, from the vertex, of the base plane onto the cutting plane. The conic is an *ellipse / parabola / hyperbola* when the vanishing line of the central collineation meets this originating circle—the circle whose image is the conic—in none / one / two points. A point $X$ is *inside / outside* a conic if it is the image of a point inside / outside the originating circle.

© Springer Nature Switzerland AG 2020
C. Baltus, *Collineations and Conic Sections*,
https://doi.org/10.1007/978-3-030-46287-1_6

A line is *tangent* to a conic when the originating line is tangent to the originating circle. Note that, generally, from a point outside a conic there are two tangents to the conic since there are two tangents to the originating circle. In one exception, both tangents to the originating circle meet that circle on the vanishing line of the central collineation; those tangents to the originating circle become *asymptotes* to the conic, a hyperbola. (Another exception is noted in an Exercise at the end of this chapter.)

La Hire's introductory lemmas concerning circles, of [5] of 1673, ended with Lemma 17; Lemmas 18–20 concern planes in 3-dimensional space, culminating in Lemma 20:

**Lemma 20** If planes in space have common line $AB$ and are cut by another plane parallel to $AB$, then the *sections* of this cutting plane with the given planes on $AB$ are mutually parallel and parallel to $AB$.

La Hire went on to prove a number of properties of conic sections. Most of the properties developed are found in the *Conics* of Apollonius, many treated as had Apollonius, others greatly simplified. Where Desargues had developed common properties of the three conic sections, La Hire treated the three separately. This work constitutes the main body of [5], in 72 pages.

## 6.1  Plani-Coniques

Then La Hire appended a 22-page section, called *Plani-coniques*, to his 1673 work. This occurred in the next year, according to La Hire's introduction (p 73); some existing copies of [5] do not include the appended section [13]. La Hire stated his goal: to demonstrate the propositions on conic sections "without imagining any solid or plane except the plane of the figure" [5, p 73]. In particular, he would not have to call on Lemma 20. He was, in fact, introducing the central collineation.

In place of pairing points, in three-dimensions, of the base plane with points of the slicing plane by a projection through the vertex of the cone, in the *Plani-coniques* La Hire gave a method of working in a single plane, in which both the figure in the base plane and the corresponding (projected) figure in the slicing plane are represented. As with perspective drawing, a figure and its projected image meet on axis $DE$, called the *formatrice*, where the slicing and base planes meet. Vanishing line $BC$, where the vertical plane meets the base plane, was called the *directice*, and must be parallel to the *formatrice*. And there is the center, called the *pole*, $A$, corresponding to the vertex of the cone, where any point, $h$, of the base plane is mapped to point $H$ in the slicing plane so that $h$, $H$, and $A$ are collinear. (Yes, in the case of Poncelet, this *directice* can be the directrix of a conic section, as we encounter in calculus.)

The construction process of the *Plani-coniques* is that of our Chap. 2, in which the center, $A$, and axis, $DE$, and vanishing line, $BC$, are given. The constructed image, $H$, of a given point $h$ must lie on the line $Ah$. See Fig. 6.1. Further, we

**Fig. 6.1** *Plani-conique*
construction of $h' = H$

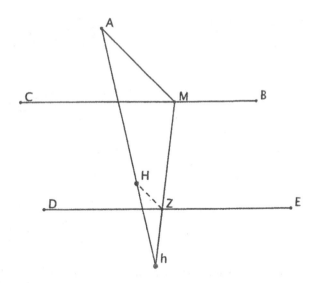

take some line on $h$ and not on $A$, meeting the *directrice* at $M$ and the *formatrice* in
point $Z$. We understand that $Z$ is fixed and $M$ must be mapped to the point at infinity
which lies on line $AM$, so line $hM$ is mapped to the line on $Z$ that is parallel to $AM$.
$H$ must be the intersection of those two lines.

In the *Plani-coniques*, after La Hire defined the construction of a corresponding
point for each given point, he showed that his construction was well-defined
(Lemma 21), that collinear points produce collinear points (Lemma 22), that lines
which meet on the *directrice* "form" parallel lines (Corollary 3 of Lemma 22), that
parallel lines generally"form" concurrent lines (Lemma 23), that tangents to the
base circle correspond to tangents to the curve "formed" by that circle (Lemma 24).
Corollary 3 of Lemma 22, namely, that lines which meet on the vanishing line
"form" parallel lines, reworks La Hire's Lemma 20 in terms of a central collineation.
Then La Hire showed that the curve formed by slicing a cone by a plane, is the curve
"formed by" a circle according to the procedure of the *Plani-coniques* [5, p 78–80].
The proof of Lemma 21 is an exercise with the Chapter Notes.

In the *Plani-coniques*, material on the conic sections already developed was
quickly revisited by the methods of the *Plani-coniques*. Some diagrams serve both
the initial 72 pages and the *Plani-coniques* addition, suggesting that the first part
was written with the second already in mind. Or, another possibility is that creating
the two-dimensional diagrams of conics in three-dimensional space pushed La Hire
to the thinking that became the *Plani-coniques* section.

The vocabulary of the *Plani-coniques* addition is new, in that where before he
had referred to line $BC$, now he calls it the *directrice*, and he asks us to look at what
had been figures representing a three-dimensional figure as simply two-dimensional.
However, the very pretty concept of a central collineation brought no new results.

**Fig. 6.2**  La Hire's *Fig.* 25,
for Lemma 17, 1673:
$H(GD, HI)$

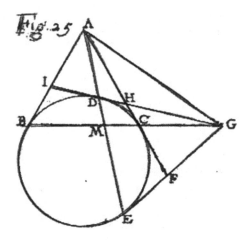

What about the *arc rampant*? Lemma 17 directs us to a solution of the *arc rampant* problem, starting with the case of a circle. See La Hire's *Fig.* 25 in our Fig. 6.2.

In Lemma 17, we have tangents from $A$ meeting the circle at $B$ and $C$, and from $G$ on $BC$ a tangent is drawn that meets the circle at $D$. The claim is that $H(GD, IH)$ when $I = BA \cap GD$ and $H = CA \cap GD$. Why? By Lemma 9, $H(GM, BC)$ and points $B, M, C, G$ are projected from $A$ to points $I, D, H, G$. The sought point of tangency on $IH$, for the *arc rampant*, is $D$.

On page 39 of [5], the result is extended to a conic—with no explicit reference to the *arc rampant* problem.

Only on the last pages of [5] does La Hire mention the *arc rampant*, when he tells us that Lemma 17 and its extension to conics

> is entirely necessary to trace *arc rampants* . . . and that is what I did and was printed in 1672 by Monsieur Bosse . . . after having seen what was done by master mason Monsieur Rouget of Aisne. I learned that Monsieur Blondel, Master of Mathematics for the Dauphin, had worked on this earlier than I had believed. I have not been able to see his work, but I have no doubt it is something beautiful since it comes from so great a man.

## 6.2   Conic Properties Developed by La Hire, 1673

What did La Hire show about conic sections in 1673? We start with several definitions.

**Definitions XII (Not from La Hire)** Let $c$ be a conic section. Two points of the plane are *conjugate* if each lies on the polar of the other; likewise, two lines are *conjugate* if each lies on the pole of the other. Most often we consider conjugate

**Fig. 6.3** $X$ and $X'$ are conjugates on line $p$ with respect to the conic

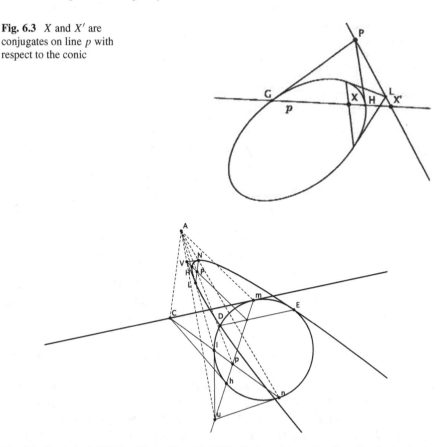

**Fig. 6.4** Parabola $LHN$. La Hire's $Fig.$ 32, 1673. Center of central collineation is $A$, $Cm$ is vanishing line

points on a given line or conjugate lines on a given point. *Conjugate diameters* are conjugate lines on the center of a conic.

In illustration, suppose $P$ is a point and $p$ the polar of $P$ with respect to a given conic. For a given point $X$ on $p$, let line $PL$ be its polar (which must lie on $P$). Then $X'$, the *conjugate* of $X$ (on line $p$), is the point at which $PL$ meets $p$. See Fig. 6.3. Here, $H(XX', GH)$.

Based on the idea that a conic section is the image of a circle under a central collineation, we summarize important results in this theorem. Its proof follows.

**Theorem 6.1 (Formation of the Conic Sections)** *Suppose $s$ is a circle mapped to conic $s'$ by central collineation $\phi$. Let $C$, on the vanishing line, have line $c$ as its polar. The center of $\phi$ is $A$ and $h$ is mapped to $H$. La Hire's Fig. 32, in our Fig. 6.4, illustrates the case when $C$ is outside the circle $s$. There $c$, the polar of $C$, is line mphu. Unlike Fig. 6.4, we will not generally assume $m$ is on the vanishing line. Then*

**Fig. 6.5** Based on La Hire's
*Fig.* 34, 1673. Center of
central collineation is $A$, $BC$
is vanishing line, $DE$ is axis.
$XY$ is the polar of $B$, so $X'Y'$
is a diameter of the hyperbola
on $D$, $E$, $X'$, $Y'$, $G'$, $P'$

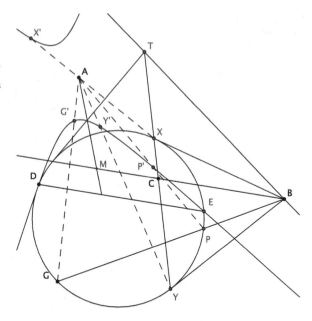

(1) *The image of c is a diameter of conic s', and the corresponding ordinates are images of the lines on C which meet circle s twice.*

(2) *Suppose C is outside circle s, as in La Hire's Fig. 32. Let u be a point on c outside the circle, with tangents from u meeting the circle s at l and n, with $p = uhm \cap Cln$. Let $\phi$ map u to V, l to L, n to N, m to M, p to P, and h to H where h is on the tangent from C. Then LN is the polar of V and is an ordinate of diameter HP of conic s'. And tangents to s' from the endpoints of ordinate LN meet the diameter (extended) at V, with $H(VP, HM)$.*

(3) *The pole of the vanishing line is mapped to the center of conic s'.*

(4) *When B and C are conjugate points on the vanishing line, then their polars map to conjugate diameters of s'.*

### Proof

(1) First consider the case that $C$ is outside the circle. See La Hire's *Fig. 32*, in our Fig. 6.4, showing the circle $s$. Let a line on $C$ meet the circle at $l$ and $n$ and meet the polar, $c$, of $C$, at point $p$. Then $H(Cp, ln)$, and since the harmonic relation is invariant under projection, then in the image we have $H(\infty P, LN)$. By La Hire's Lemma 3, $P$ is the midpoint of chord $LN$ of the conic $s'$. This holds for any line on $C$; further, since $C$ is mapped to infinity, then all the chords $LN$ are parallel. This means that the image of line $c$ is a diameter of the conic $s'$, and the collection of chords $LN$ are the corresponding ordinates.

(2) The claim about a pole and its polar and the related tangents holds for the circle $s$ and that property is preserved under projection $\phi$.

(3) The polars of the points of the vanishing line all lie on the pole, $x$, of the vanishing line. Those polars are mapped by $\phi$ to diameters, so $X$, the image of $x$, lies on all the diameters of conic $s'$. So $X$ is the center of $s'$.

(4) Since $B$ and $C$ are conjugate points, they each lie on the polar of the other. Those polars of points on the vanishing line are mapped to diameters, and conjugacy is preserved under projection, so the two diameters are conjugates.

□

In Fig. 6.5, we see how this theorem plays out in the case of a hyperbola. Point $B$, on the vanishing line $BC$, has polar $XY$, whose image is diameter (not drawn) $X'Y'$ joining points of opposite branches of the hyperbola. Here $C$ is on the vanishing line inside the circle, so the polar of $C$—which is $BT$ in Fig. 6.5—lies outside the circle. By a corresponding argument, the chords on $C$, such as $XY$, are mapped to parallel ordinates, which are, again, bisected by the image of the point, such as $T$, where the chord meets the polar of $C$.

The pole (not drawn), $K$, of the vanishing line is mapped to the center of conic $s'$ since $K$ lies on the polars of all the points on vanishing line $BC$, meaning that $K'$ is on all diameters of $s'$.

The ellipse, parabola, and hyperbola are distinguished by the number of points of the base circle mapped to infinity: none for the ellipse, which is bounded; one for the parabola, which goes to infinity in one direction, and two for the hyperbola, which approaches infinity along the two distinct asymptotes.

For the case of the ellipse, see the three-dimensional sketch of a cone, Fig. 6.6, whose vertical plane meets the base plane in vanishing line $BC$ outside the base circle $hfnl$.

The plane of the base circle and of the slicing plane are represented separately in Fig. 6.7. The conic is an ellipse since $BC$ is completely outside the base circle. We see that $B$ and $C$ are conjugate points on vanishing line $BC$, so they give rise to conjugate diameters $HN$ and $FL$.

Numerous properties of conic sections are developed in the *Conics* of Apollonius and in the books on conic sections which have followed on that work. We treat just a few of the most important and prettiest following the methods of La Hire.

**Theorem 6.2 (Example 1 La Hire [5, p 29])** *Let $V$ be a point outside a given conic and let the tangents from $V$ meet the conic in points $N$ and $L$, and let $P$ be the midpoint of $NL$. Then $VP$ is a diameter of the conic section, with ordinates parallel to $NL$. [Apollonius Book 2 Prop. 29]*

***Proof*** Figure 6.8 is the base circle which is projected to the conic on the slicing plane, with $V$, $L$, and $N$ the projections of $u$, $l$ and $n$, respectively. Let line $ln$ meet the vanishing line (dotted) at $C$. The *polar* of $C$, meeting the circle twice, must lie on $u$ by La Hire's Lemma 10, and it meets line $ln$ at $p$ so $H(Cp, ln)$. So projected to the slicing plane, $H(\infty P, LN)$, which means $p$ is projected to the midpoint of $LN$, and all the lines on $C$ in the base plane are projected to lines parallel to $LN$ and meet the conic in chords bisected by $VP$. So $VP$ is a diameter.                    □

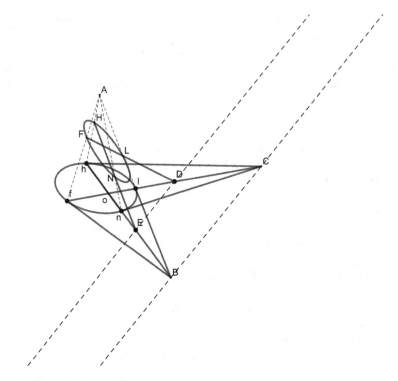

**Fig. 6.6** Cone sliced in an ellipse

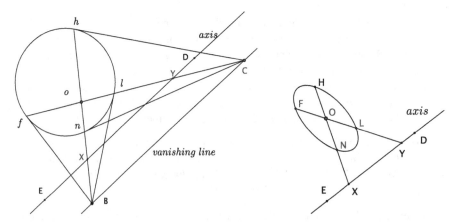

**Fig. 6.7** Left: Base of cone for ellipse. Right: Cone slice for ellipse

This theorem implies that on any point outside a given conic there is a diameter of the conic. If a point $V$ is on or inside a conic which has center $O$, take any point on line $VO$ that lies outside the conic and apply the procedure of the proof above.

**Fig. 6.8**  La Hire [5] p 29,
Apollonius Book 2 Prop. 29

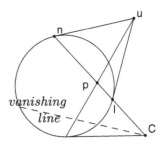

**Example 2  Parabola.** Let's look at La Hire's *Fig.* 32 of 1673 in our Fig. 6.4. *lhnm* is the base circle. As before, the central collineation $\phi$ has center $A$ and maps $h$ to $H$. This time the vanishing line is $Cm$, tangent to the base circle at $m$. Therefore, the conic will be a parabola, $LHN$. *Fig.* 32 served La Hire both in his initial consideration of conics as sections of a conic surface in three-dimensions, and, again, in the *Plani-coniques* section where the given circle and its conic image lie in the same plane.

The slicing plane meets the base circle in the axis, which must be parallel to the vanishing line and on $VP \cap up$. The resulting parabola looks a bit awkward, but the curious reader is encouraged to follow the mapping procedure of a central collineation to verify that the diagram was carefully made following that procedure.

The diameters of the parabola will be parallel since they are images of the chords on $m$, on the vanishing line.

Further, consider a tangent to the parabola at $L$, as in La Hire's *Fig.* 32. Since $H(up, hm)$, then by projection from $A$, $H(VP, H\infty)$. So $H$ is the midpoint of $VP$. Further, the tangent at $H$ is parallel to ordinate $LN$, so the tangent at $H$ bisects $LV$.

To summarize:

**Theorem 6.3 (Parabola Properties)**

*a. The diameters of a parabola are parallel.*
*b. Let $HP$ be a diameter of a parabola, with its vertex at $H$, and let the tangent to the parabola at a point $L$ meet this diameter extended at $V$, and let $LP$ be the corresponding ordinate. Then vertex $H$ is the midpoint of $VP$, and the tangent at $H$ bisects $LV$.*

**Example 3  Hyperbola.** Here is a pretty property of the hyperbola, proved by Apollonius in Book 2 Prop. 8. The projection of the base circle onto the slicing plane will be a hyperbola when the vanishing line, $BC$, meets the base circle twice, at points $n$ and $h$ in Fig. 6.9 Right. Since $n$ and $h$ are mapped to infinity, the tangents at $n$ and $h$ are mapped to asymptotes to the conic section. We consider a branch of a hyperbola with center $O$, and a line which cuts the branch of the hyperbola at $P$ and $G$, and cuts the asymptotes at $V$ and $W$, shown in Fig. 6.9 Left. We'll show: $PW = GV$.

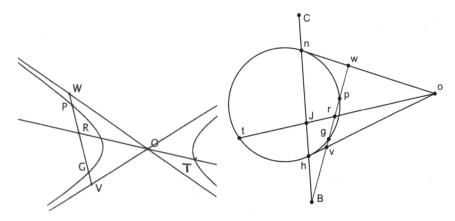

**Fig. 6.9** Left: Hyperbola theorem. Top Center: Base plane for Hyperbola theorem

The corresponding figure in the base plane is in Fig. 6.9 Right. Line $PG$ cutting the hyperbola corresponds to line $pg$ in the base plane. Let $pg$ meet the vanishing line at $B$. $o$, the pole of $BC$, is mapped to $O$, the center of the hyperbola. Draw line $ot$, the polar of $B$, meeting $BC$ in $j$ and $pg$ in $r$. As we have seen, $OT$, the image of the polar of $B$, is a diameter of the hyperbola, and $PG$ will be a corresponding ordinate. Now by La Hire's Lemma 9, $H(Br, gp)$. Also, $H(Bj, hn)$, and so, by projection from $o$, $H(Br, vw)$. Then since $B$ is mapped to infinity, $R$ is the midpoint of $PG$, and $R$ is the midpoint of $VW$, so in the slicing plane, $PW = GV$ [5, p 27]. □

Again, to summarize in a theorem:

**Theorem 6.4** *Suppose a line cuts one branch of a hyperbola at P and G and cuts the asymptotes to that branch at V and W. Then $PW = GV$.*

The biggest part of the material in Book 1 of Apollonius's *Conics* has been covered in this chapter. Especially lengthy for Apollonius was the argument that all lines on the center are also diameters, after the principal diameter was found early in the book. La Hire's treatment is much shorter and simpler. Desargues' work was briefer still. A second simplification over the methods of Apollonius, found in La Hire's work, and, indirectly, in that of Desargues, involves tangents to the conic sections. For one thing, Apollonius used a more demanding definition of a tangent. He always showed, by a proof by contradiction, that the proposed tangent could not meet the conic in a second point. La Hire had it easier: a tangent to a conic was just the image of a tangent to the base circle, and Euclid's Book 3 told us what we need to know about a tangent to a circle.

Example 4, which follows, is based on La Hire's *Fig.* 52 of 1673, and the corresponding *Sixiemème Partie*. The figure is for the case of an ellipse. It is one of several examples in which La Hire demonstrated the pointwise construction of a conic, a topic without interest to Apollonius.

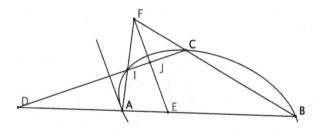

**Fig. 6.10** Based on La Hire [5] *Fig.* 52

**Example 4  Ellipse.** Problem: Given a diameter $AB$ of an ellipse, with the tangent at $A$, and given another point, $C$, of the ellipse, construct various other points of the ellipse.

Solution: See Fig. 6.10. Our plan is take a point $D$ on diameter $AB$ extended, and then construct a corresponding point $I$ on the ellipse.

Draw $E$ on $AB$ so $D$ and $E$ are harmonic conjugates with respect to $A$ and $B$. Draw $CD$ and line $BC$, and at $E$ draw a line parallel to the tangent at $A$, meeting $CB$ at $F$. Let $I$ be $DC \cap AF$.

Think of $DC$ as a secant to the ellipse, meeting the ellipse at $C$ and a point $I^*$. So $H(DJ, I^*C)$ where $J$ is on the polar of $D$. We constructed $D$ and $E$ so $E$ is on the polar of $D$, so $H(DE, AB)$.

Since $EF$ is parallel to the tangent at vertex $A$, then $EF$ is an ordinate with respect to diameter $AB$. So $EF$ is the polar of $D$, which means that $J$ is on $EF$.

Since $H(DJ, I^*C)$ and $H(DE, AB)$, harmonic sets with a common point $D$, then the lines joining the other corresponding points of the harmonic sets are concurrent, at $F$. So $I^* = I$. This mean the constructed point, $I$, lies on the ellipse.                                                                          □

We note, as a theorem, the related tangent property whose proof required much labor by Apollonius, in Book 1 Prop 34, 36. The property is also illustrated in Fig. 6.10 when $K$ is the intersection of the conic and $EF$.

**Theorem 6.5**  *If a tangent at a point $K$ of a conic meets a diameter $AB$ (extended) at $D$, then $H(DE, AB)$ where $EK$ is the ordinate on $K$.*

## 6.3  Notes and Exercises

1. Although the concept of *pole* and *polar* goes back to, at least, Apollonius, we only see terminology in the work of Desargues, and the modern *pole-polar* terminology from 1811, when *pole* appeared in a paper by Servois, *Annales de Mathématiques Pures et Appliquées* Vol 1 p 337, and *polar* the next year, Vol 3 p 297. [12, Steiner]]

2. In the preface to [6], of 1679, La Hire looked back on the 1673 work and the *Plani-coniques*.

> Several years ago I had printed a Treatise of Conic Sections [1673] by a new method, where I demonstrated their principal properties in the cone: but those who are not accustomed to demonstrations by the intersections of planes and solids had difficulty understanding them, even though they are very simple once one comprehends them. This occasioned me to seek another method, where by simply describing these curves in a plane, without use of a cone, I might demonstrate the same properties as with the solid; and after trying the [algebraic] method which I use here ... I abandoned it, not being able to surmount all the difficulties I encountered, and I contented myself with reducing the cone and the sections onto a plane, which I called *Planiconiques* ....
>
> But although very advantageous in pleasing the scholars, one should not make it the principal object of study, and entirely neglect the instruction of those who want to learn, and I believe they should be happy when they are given different routes to the same goal, so each can choose a way according to his inclination and the capacity of his genius.

3. La Hire's [7] of 1685 does not mention *Plani-coniques* and has less emphasis on central collineations than the 1673 work. However, it is worth mentioning one innovation over the 1673 work, something that had appeared in Desargues' [4]. That addition is the dual of a harmonic set of lines, which La Hire called a *harmonicale*: four concurrent lines where any line meeting those four does so in a harmonic set. We'll call it a *harmonic pencil*; we'll see this concept further exploited in the work of Poncelet.

4. In the eighteenth century, La Hire's work of 1673 was little known. His 1685 work, on the other hand, was known to those interested in conic sections. D'Alembert, for example, recommended La Hire's 1685 work on conic sections in his article *Conique* in the 1751 *Encyclopédie*: there "one finds the properties of conic sections *traitées fort*." [1]. In the late eighteenth century, La Hire may have been better known for his work related to the construction of vaults. "His *Traité de méchanique* (1695) and a memoire of 1712, 'Sur la construction des voûtes dans les édifices' contain a first precise formulation of the problems of the egg and the wall." (Problem of the egg: "How can an apparently fragile curve support heavy loads?") [2, p 321–325].

5. The one comprehensive study of La Hire's projective geometry is a PhD thesis by Zbynek Sir [11], available on the internet. Sir examined what appears to be all the biographic information available on La Hire. His careful examination of La Hire's work includes a discussion in which he concludes that Desargues' influence on La Hire was minimal.

6. The pointwise construction of conics, as in Example 4, was of little interest to Apollonius but drew the attention of some in the seventeenth century. Mydorge devoted Book II of [10] to such constructions. Bosse, interested in practical applications, suggested workers plot points with pegs on a curve they wished to trace and then fit a mechanical spline to the pegs.

7. Luigi Maierù's [8], in Italian, is a good survey of seventeenth century work on conic sections.

**Exercise 1** The points of tangency from an outside point to a conic can be constructed by just a straightedge, following La Hire's Lemma 10. In Fig. 6.11 Left, complete the construction of points of tangency to the conic of tangents drawn from $A$, by straightedge only.

**Exercise 2** Suppose a circle is mapped to a conic by a central collineation and that from a point $A$ outside that circle exactly one tangent from $A$, namely line $m$, is not the vanishing line of the central collineation but meets the circle at a point on the vanishing line. What shape is the resulting conic and how is the image of $m$ related to it?

**Exercise 3 (You May Draw Parallels and Tangents "By Eye")** In Fig. 6.11 Right, why is the image of the circle under the central collineation, with the given axis, vanishing line, and center $A$, a hyperbola? Construct by straightedge and by parallels (1) the asymptotes of the hyperbola, (2) the image of point $X$, (3) the diameter of the hyperbola that lies on the image of $X$.

**Exercise 4** In the given parabola, in Fig. 6.12 Left, construct a diameter. Then construct the axis (the diameter which is perpendicular to its ordinates). You may "construct" parallels by eye or with a device.

**Exercise 5** In the given diagram, in Fig. 6.12 Right, central collineation $\phi$ has center $A$ and its vanishing line is a tangent to the base circle at $m$, axis $DE$, and center $A$.

(1) Draw a diameter of the parabola which is the image of the circle under the central collineation. (2) In the base plane, construct the line which is mapped to the axis of the parabola. For a challenging exercise, do not use Exercise 4. Hint: Where are the lines of the base plane that are mapped to the ordinates corresponding to the axis of the parabola?

**Exercise 6 (Proof of La Hire's Lemma 21 of [5])** Referring to Fig. 6.1, we are given center $A$, vanishing line $BC$, and axis $DE$. Show that for a given point $h$ we

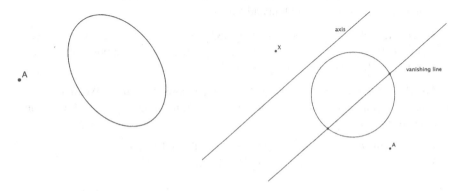

**Fig. 6.11** Left: Exercise 1. Right: Exercise 3

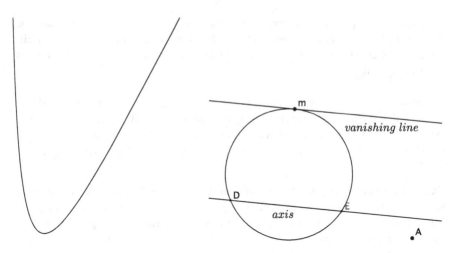

**Fig. 6.12**  Left: Exercise 4. Right: Exercise 5

**Fig. 6.13**  Seydewitz's
theorem, Exercise 7

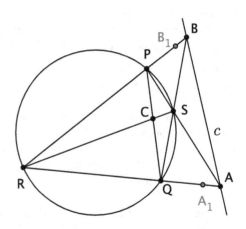

arrive at the same image $H$ no matter how, in the construction process, we select a
line on $h$ to meet the vanishing line and axis.

**Exercise 7 (Seydewitz's Theorem [3, p 253])** Let $s$ be a conic with inscribed
triangle $PQR$. Take a point $C$ on side $PQ$, and let $c$ denote the polar of $C$. Then
sides $RP$ and $RQ$ meet $c$ in conjugate points.

See Fig. 6.13. For the proof, we can assume the conic is a circle. We first let line
$RC$ meet the conic in $S$. We draw lines $PS$ and $QS$, meeting $RQ$ and $RP$ in points
$A_1$ and $B_1$, respectively.

$a$. Explain why $B_1C$ is the polar of $A_1$ and $A_1C$ the polar of $B_1$.
$b$. Explain why $A_1$ and $B_1$ are on the polar of $C$.

## 6.4 Some Hints and Solutions

### Exercise 1
Solution: If a secant from $A$ meets the conic at $B$ and $C$, and a second secant meets the conic at $E$ and $F$, then $BE \cap CD$ and $BF \cap CE$ lie on the polar of $A$, and the polar meets the conic at the points of tangency.

### Exercise 2
Partial solution: The vanishing line will meet the circle in two points, so the image of the circle is a hyperbola.

### Exercise 3
Hint: Once one diameter is found, you will need to find the diameter whose ordinates are perpendicular to that diameter.

### Exercise 4
Hint for the second part: Take $C$ on the vanishing line so $CA \perp Am$.

### Exercise 5
Solution. Refer to Fig. 6.1. A second line on $h$ is supposed, meeting $DE$ and $CB$ at $Y$ and $U$, respectively. The generated image of $h$, denoted $L$, satisfies, by parallels,

$$\frac{Ah}{AL} = \frac{Uh}{UY} = \frac{Mh}{MZ} = \frac{Ah}{AH}, \text{ so } H = L.$$

### Exercise 6
Hint for $a$. Consider quadrilateral $QSPR$.
Solution to $b$. Since $C$ lies on the polars of $A_1$ and of $B_1$, then $c$ lies on $A_1$ and $B_1$, so $A_1 = A$ and $B_1 = B$. As $A$ and $B$ each lie on the conjugate of the other, they are conjugates.

## References

1. Jean Le Rond d'Alembert, Conique, in *Encyclopédie ou Dictionnaire Raisonné*, Paris: Braisson et al., 1751.
2. Edoardo Benevenuto, *Introduction to the History of Structural Mechanics: Part II: Vaulted Structures and Elastic Systems*, New York: Springer, 2012.
3. H. S. M. Coxeter, *Introduction to Geometry*, New York: Wiley, 1969.
4. Girard Desargues, *Brouillon project d'une atteinte aux événements des rencontres d'un cône avec un plan*, in [9] (trans. Field) and original http://gallica.bnf.fr/ark:/12148/bpt6k105071b/f1.imageParis, 1639.
5. Philippe de La Hire, *Nouvelle Méthode en Géométrie pour les Sections des Superficies coniques et Cylindriques*, Paris, 1673.
6. Philippe de La Hire, *Nouveaux Elements des Sections Coniques; Les Lieux Géométriques; La Construction, ou Effection des Équations*, Paris: André Pralard, 1679; Reissued: Whitefish, MT: Kessinger Publishing, 2009.

7. Philippe de La Hire, *Sectiones Conicae en novem libros distributae*, Paris 1685; French translation by Jean Peyroux, *Grand Livre des Sections Coniques*, Paris: Blanchard, 1995.

8. Luigi Maierù, *Le sezioni coniche nel Seicento*, Catanzaro, Italy: Rubbetino, 2009.

9. J. V. Field and J. J. Gray, *The Geometrical Work of Girard Desargues*, New York: Springer, 1987.

10. Claude Mydorge, *Prodromi catoptricorum et dioptricorum: sive conicorum operis ad abdita radii reflexi et refracti mysteria praevii et facem praeferetis*, Books 1 and 2, 1631, Books 3 and 4, 1639.

11. Sír, Zbynek, *Les sections coniques chez Philippe de La Hire*. Ph.D. Thesis, Université Pierre et Marie Curie (Paris VI), 2002. http://www.karlin.mff.cuni.cz/~sir/papers/These.pdf

12. Jacob Steiner, *Die geometrischen Konstructionen, ausgeführt mittelst der geraden Linie und Eines festen Kreises, ...*, Berlin: Dümmler, 1833, translated and edited by M. E. Stark and R. C. Archibald, as *Jacob Steiner's Geometrical Constructions with a Ruler*, New York: Scripta Mathematica, 1950.

13. H. Wieleitner, Über die 'Plani-coniques' von de La Hire, *Archiv für die Geschichte der Naturalwissenschaften und der Technik* Vol. 5, 1913, Leipzig, 49–55.

# Chapter 7
# Central Collineations: Complete Quadrilateral, Involution, and Hexagon Theorems

## 7.1 The Complete Quadrilateral

An alternate definition of a harmonic set, based on the *complete quadrilateral*, emerged in the nineteenth century, a definition that does not depend on segment length. The concept appeared in the work of Desargues, although the name was first used by L. Carnot in 1803 [2, p. 275]. That it produces a harmonic set was recognized by La Hire in 1685 [5, Prop XX]. Its dual, the *complete quadrangle*, is a different way of viewing the same figure. (In the period we cover, the figure was always called a complete quadrilateral.)

**Definition XIII** If points $J$, $V$, $K$, $E$ have no three collinear, then $JVKE$ is a *complete quadrangle*, and the four sides form a *complete quadrilateral jvks*. See Fig. 7.1.

**Theorem 7.1**

(1) *Let $JVKE$ be a complete quadrilateral. Let $JE$ meet opposite side $VK$ at $B$, and $JV$ meet opposite side $EK$ at $D$, and let $BD$ meet diagonal $VE$ at $C$ and meet diagonal $JK$ at $A$. Then $\{A, B, C, D\}$ is a* harmonic set, *with $A$ and $C$ harmonic conjugates of $B$ and $D$.*
(2) *On the other hand, given a harmonic set where $H(AC, BD)$, then there is a complete quadrilateral $JVKE$ producing harmonic set $\{A, B, C, D\}$ as in (1).*

We begin the proof with a Lemma:

**Lemma** *Given four collinear points, $A$, $A'$, $B$, $B'$, and a given line $l$ on none of the four points, there is a central collineation, with axis $l$, that maps $A$ to $A'$ and $B$ to $B'$.*

***Proof of the Lemma*** We can define the central collineation by finding the center $X$ on line $AB$ so $A$ is mapped to $A'$ and $B$ to $B'$. To do this pick two points, $R$

© Springer Nature Switzerland AG 2020
C. Baltus, *Collineations and Conic Sections*,
https://doi.org/10.1007/978-3-030-46287-1_7

**Fig. 7.1** Complete
Quadrilateral $VKEJ$:
$H(AC, BD)$

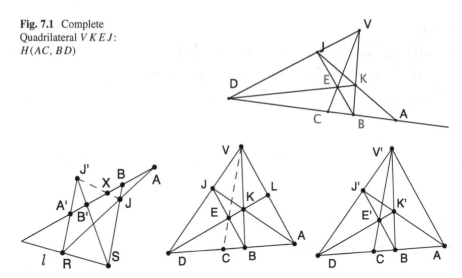

**Fig. 7.2** Left: Lemma. Center and Right: Theorem 7.1

and $S$ on $l$—and not on $AB$. See Fig. 7.2 Left. Draw $AR$, $RA'$, $BS$ and $SB'$. Let $J = AR \cap BS$ and $J' = A'R \cap B'S$. Then $X$ will be $AB \cap JJ'$ and the $(X, l)$ collineation that maps $A$ to $A'$ also maps $B$ to $B'$.

***Proof of the Theorem*** We first prove part (2). Suppose points $A$, $B$, $C$, $D$ are collinear where $H(AC, BD)$ by the ratio definition. This means $AB \cdot CD = AD \cdot BC$ and $A$ and $C$ separate $B$ and $D$. See Fig. 7.2 Center. We create an isosceles triangle $DAV$, with $VD = AD$, $J$ on $VD$ so $VJ = AB$. It follows that $AV$ and $BJ$ are parallel lines. Let $E$ and $L$ be midpoints, respectively, of $BJ$ and $AV$. Then $LD$, $AJ$, and $VB$ will be concurrent, at $K$. By similar triangles, (as constructed in La Hire's 1673 Lemma 1),

$$\frac{EK}{KL} = \frac{EB}{VL} = \frac{JE}{VL} = \frac{DE}{DL} \quad \text{so} \quad ED \cdot KL = EK \cdot DL.$$

This means $H(DK, EL)$. By La Hire's Lemma 7, the two harmonic sets are in perspective, from $V$. So $VE$ lies on $C$ [4]. So complete quadrilateral $VKEJ$ produces the harmonic set $\{A, B, C, D\}$.

Conversely, suppose complete quadrilateral $JEKV$ gives set $A, B, C, D$ on line $l = BD$: opposite sides $VK$ and $JE$ meet at $B$, $JV$ and $EK$ meet at $D$, $JK \cap l = A$, $VE \cap l = C$. We'll show $AD \cdot BC = AB \cdot CD$. The plan is to apply a central collineation, with axis $l = AD$, which gives the same sort of figure as in the proof of the first part. On ray $DV$, place points $J'$ and $V'$ so, in length, $DB = DJ'$ and $DA = DV'$. By the lemma above, there is a central collineation with axis $AD$ and so $J$ is mapped to $J'$ and $V$ is mapped to $V'$. The resulting figure—See Fig. 7.2 Right—has $BJ' \parallel AV'$. The mapping is a central collineation with axis $l$, so $V'E'$

is on $C$, etc. As in the first part of the proof, we conclude that $\{A, B, C, D\}$ is a *harmonic set*, with $AD \cdot BC = AB \cdot CD$. $\square$

**Corollary (Invariance of the Harmonic Relation)** *A collineation maps a harmonic set to a harmonic set since a complete quadrilateral is mapped to a complete quadrilateral.*

## 7.2 Involution

The term *involution* was introduced by Girard Desargues in his *Brouillon* of 1639. That work, which seems to have disappeared by the time of Desargues' death in 1661, was only recovered in 1845 in a transcript by Philippe de la Hire. But enough information about involution was available to Poncelet that he could revive the concept, still under the name *involution*, in his 1822 *Traité*. Desargues described a particular pairing of collinear points, generally involving three pairs, although he treated a four-point involution, later known as a *harmonic set*, and allowed for any number of pairs to be included in an involution. We will say more about Desargues' work in a later chapter.

And we will later give a more general definition of *involution*, but for now we define a particular case, which we call an *involution collineation*.

**Definitions XIV** An *involution collineation* of a line $m$ is a $(V, l)$ collineation $f$ restricted to line $m$ where $V$ lies on $m$ with the following property:

- for all $x$ on $m$, $f(x) = y$ exactly when $f(y) = x$.

We know that for given points $V$ and $L$, we can define a unique central collineation with center $V$ by designating a line $l$ on $L$, with $l \neq VL$, to serve as the axis and selecting any two other points on $VL$, $A$ and $A'$, where $A'$ is to be the image of $A$. We will show that any other $B$ on $VL$ is mapped to $B'$ where $B'$ is mapped to $B$. And that $(V, l)$ collineation will effect an involution collineation on any other line on $V$.

The properties are in the following theorem.

**Theorem 7.2 (Involution Collineation Theorem)**

a. *Given a line on points $V$ and $L$, there is a $(V, l)$ collineation, for an axis $l$ on $L$, mapping any $A$ on $VL$ to a point $A'$ so $A'$ is mapped to $A$. Such a collineation is defined to be an* involution collineation *of line $VL$.*

b. *All other points $D$ of $VL$ are interchanged with their images $D'$ under this $(V, l)$ collineation.*

c. *The same involution collineation is an* involution collineation *on any other line on $V$.*

d. *An involution collineation with center $V$ and fixed point $L$ maps any point $A$ of $VL$ to its harmonic conjugate with respect to points $V$ and $L$.*

**Fig. 7.3**  Theorem 7.2,
Involution that interchanges
$A$ and $A'$

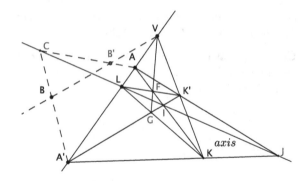

***Proof***  See Fig. 7.3. For *a.*, we first construct point $A'$ on $VL$ so $H(AA', VL)$. In other words, $A'$ is to be the harmonic conjugate of $A$ with respect to $V$ and $L$. That construction can be achieved by constructing complete quadrilateral $FK'KG$. (We can take any $K$ and $K'$ collinear with $V$, and not on line $VL$, and then complete the construction as pictured.) Let $AK$ meet $A'K'$ at $I$. Our involution collineation will have center $V$, axis $l = LI$, and map $A$ to $A'$. By the corollary above, *Invariance of the Harmonic Relation*, with $V$ and $L$ fixed and $A$ mapped to $A'$, then $A'$ must be mapped to its harmonic conjugate with respect to $V$ and $L$, namely, $A$.

For *b.*, *c.*, and *d.*, take any other line $VB$ on $V$. Let line $A'B$ meet the axis at $C$. We draw $AC$, meeting line $VB$ at a point that must be $B'$ since $A'$ is mapped to $A$. But since, also, $A$ is mapped to $A'$, $B'$ must be mapped to $B$. And if we take any point $D$ on line $VL$, using the argument above, $D$ and $D'$ are interchanged by the same central collineation. We know from the work of La Hire, and by the corollary to Theorem 7.1, that a harmonic set on one line is projected to a harmonic set on another line, so by projection from $C$, $B$ and $B'$ are harmonic conjugates with respect to $V$ and the point at which $VB$ meets the axis.                    □

## 7.3   Collineations that Map a Circle to a Circle

We first note a central collineation, not the identity, that maps a circle to itself.

**Theorem 7.3** *Given a circle and a point $V$ outside the circle, let $V$ be the center and let the polar of $V$, $l$, be the axis, and we define collineation $\phi$ as the $(V, l)$ collineation that maps $A$ to $B$, when the diameter on $V$ meets the circle at $A$ and $B$. (Let $AB$ meet $l$ at $L$.) Then the circle is mapped to itself.*

***Proof***  See Fig. 7.4. We know that $H(VL, AB)$. Since $V$ and $L$ are fixed and a harmonic set is mapped to a harmonic set by a collineation, then $B$ is mapped to $A$ by the same collineation. So we have an involution collineation. Let another line on $V$ meet the circle at $G$ and $H$ and meet axis $l$ at $M$. By La Hire's Lemma 9,

**Fig. 7.4** Theorem 7.3: Circle
mapped to itself

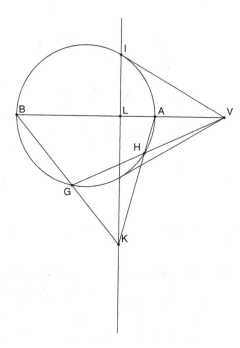

$H(VM, HG)$, so the involution collineation exchanges points $H$ and $G$. $H$ is
arbitrary, so the circle is mapped to itself.                                    □

We restate and reprove the *Pole-Polar Theorem*.

**Theorem 7.4 (Pole-Polar Theorem)** *Given a circle, a point $X$ and its polar $x$.
Then a point $Y$ is on line $x$ exactly when $y$, the polar of $Y$, is on $X$.*

*Proof*  First, we'll consider the case when $X$ and $Y$ are both outside the circle, with
$Y$ on $x$, as in Fig. 7.5 Left. We'll show $y$ is on $X$. Draw the tangent $YG$ to the circle.
Let the diameter on $X$ meet the circle in $A$ and $B$, and let $\phi$ be the $(X, x)$ involution
collineation that maps $A$ to $B$. $G$ is mapped to a point $H$ on the circle so the second
tangent from $Y$ meets the circle at $H$. A point and its image are collinear with the
center, so $GH$, the *polar* of $Y$, is on $X$.

And if we suppose that line $y$ is on $X$, and $y$ meets the circle twice, at $G$ and at
$H$, then the tangents at $G$ and $H$, which meet at $Y$, are images of each other, so they
meet on the axis, $x$.

Now suppose $X$ is inside the circle, but not the center, and $Y$ is outside the circle.
See Fig. 7.5 Right. So $x$, the *polar* of $X$, is outside the circle, meeting the diameter
$AB$ on $X$ at the point $L$ where $H(LX, AB)$. Suppose $Y$ is on $x$. So $Y$ is fixed
in the central collineation with center $X$ and axis $x$. To find $y$, we draw a tangent
$YP$ meeting the circle at $P$. Since a collineation maps a tangent to a tangent, $P$ is
mapped to $Q$ where $YQ$ is the other tangent to the circle. But $X$ is the center of the
collineation, so $P$ and $Q$ and $X$ are collinear. So $y$, which is $PQ$, is on $X$. On the

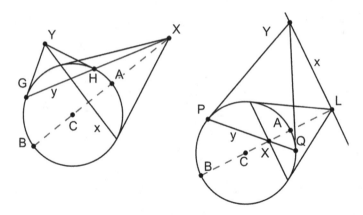

**Fig. 7.5**  Theorem 7.4, $Y$ is on $x$ iff $y$ (polar) is on $X$

other hand, suppose $y$ is on $X$, meeting the circle at $P$ and $Q$. Draw the tangent at $P$, meeting $x$ at a point $T$. Since $P$, $Q$, and $X$ are collinear and a tangent is mapped to a tangent, then $Q$ is $P'$ by the same central collineation. This means the tangent at $Q$ also meets $x$ at $T$, which means $T$ must be $Y$, the *pole* of $y$.

If $X$ and $Y$ are both inside the circle, then $x$ is outside the circle and we cannot have $Y$ on $x$ or $y$ on $X$.

The cases where $X$ is on the circle or at the center of the circle are easily handled.
$\square$

An important theorem says we can map a circle to a circle by a central collineation in which we may choose the center of the image circle inside the original circle or we may choose the vanishing line of the central collineation. The second form, when a given line is designated as the vanishing line, was important to Poncelet.

**Theorem 7.5 (Circle-to-Circle Theorem (See Poncelet, [8, Art. 110, 120]))** *Suppose we have a circle $c$ with interior point $X$, not the center, and line $TH$ outside the circle that is the polar of $X$.*

*Then there is a central collineation with vanishing line $TH$ mapping $c$ to a circle, $c'$, with center $X$.*

***Proof*** See Fig. 7.6. Let the line on $X$ that is parallel to $TH$ be the axis for the central collineation we seek, meeting the circle at $R$ and $E$. Let the diameter on $X$ meet the given circle at $A$ and $B$ and meet the vanishing line at $T$. To construct the center, $V$, of the collineation, first take $B'$ on the diameter on $A$ so $B'X$ equals $XR$. Let $BE$ meet the vanishing line at $H$. Let the line on $H$ that is parallel to $B'E$ meet $AB$ at $V$; $V$ is to be the center of the central collineation and $ER$ its axis. The resulting central collineation maps $B$ to the constructed $B'$ since the image of line $BE$ is on $E$ and parallel to $VH$. The image of $c$ is the circle with center $X$ and radius $XR$.

**Fig. 7.6** Circle-to-Circle
Theorem, mapping to a circle
with center $X$

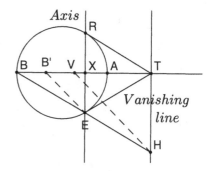

Why? The image is an ellipse since the vanishing line does not meet the given circle. $RE$ will be a diameter of the image conic since its *pole*, $T$, is mapped to infinity, and the midpoint, $X$, of that diameter will be the center of the image conic. Segment $XB'$ will be half the conjugate diameter since $T$ and infinity are conjugate points on $TH$ with respect to the given circle, and so are mapped to conjugate diameters. Recall that $XB' = XR$; an ellipse with equal perpendicular conjugate diameters must be a circle.                                                    □

## 7.4 Theorems of Pascal and Brianchon

Blaise Pascal (1623–1662), who accompanied his father to gatherings of the Paris circle about Father Marin Mersenne (1588–1646), produced a short essay in 1640. The title was *Essay pour les Coniques*; Pascal was 16 years old. In the essay, he used terminology and methods which he learned from Desargues [6]. Among the theorems stated in the essay was one we are calling Pascal's Hexagon Theorem, although Pascal's 1640 theorem was in a different form and without proof. It appears he first found the theorem for a circle, from which it immediately follows for a conic. Here is the statement of Pascal's Hexagon Theorem. Figure 7.7 Left serves to illustrate both Pascal's Hexagon Theorem and its dual, Brianchon's Theorem, in the case when the conic is a circle.

**Theorem 7.6 (Pascal's Hexagon Theorem)** *Given points B, C, D, E, F, G on a circle, then opposite sides of hexagon BCDEFG meet in collinear points. Since a conic is the image of a circle under a central collineation, the same is true when the six points are on a conic.*

The dual of Pascal's Hexagon Theorem, Brianchon's Theorem, was proved by Charles Julien Brianchon (1783–1864), in 1806 [1]. His paper gives no indication of awareness of the concept of dual statements. Brianchon used Pascal's Theorem in his proof. We, however, will first prove Brianchon's Theorem, from which Pascal's Theorem follows.

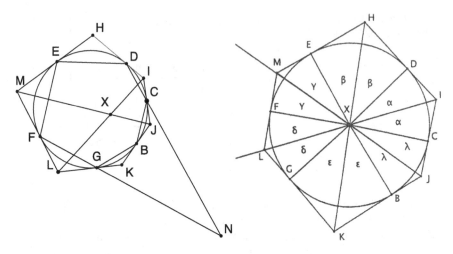

**Fig. 7.7**  Proof of Brianchon's Theorem

**Theorem 7.7 (Brianchon's Theorem)** *If a hexagon is circumscribed about a conic, then the opposite vertices of the hexagon are joined by concurrent lines.*

***Proof of Brianchon's Theorem*** The given conic is the image of a circle under a central collineation, so we prove the theorem when we are given a hexagon circumscribed about a circle, with vertices $H$, $I$, $J$, $K$, $L$, and $M$. See Fig. 7.7 Left. We need to show that $HK$, $IL$ and $JM$ are concurrent. We let $IL$ and $JM$ meet at $X$, and we need to show $H$, $X$, and $K$ are collinear. $X$ lies inside the circle for convex $HIJKLM$. By the Circle-to-Circle Theorem, we map the circle by a central collineation to another circle with center $X$. Keeping the original names for the image points, diagonals $IL$ and $JM$ meet at the center, $X$.

We now look at a circle with center $X$, in Fig. 7.7 Right. The sides of the circumscribed hexagon are tangent to the circle, so joining the hexagon vertices to the center and the points of tangency to the center, we have angles in congruent pairs; $\alpha = \alpha$, $\beta = \beta$, etc. Because $JM$ and $LI$ meet at $X$, then $\alpha + \lambda = \delta + \gamma$. Since $JM$ is a diameter, $\gamma + 2\beta + 2\alpha + \lambda = \gamma + 2\delta + 2\epsilon + \lambda$. Therefore, $\beta + \alpha = \delta + \epsilon$, and this means $H$, $X$, and $K$ are collinear. Finally, since the theorem holds for a circle and projection is one-to-one and preserves collinearity, it holds for a conic section.                                                                                                 □

Now, Brianchon's Theorem implies Pascal's Theorem. We wish to show that opposite sides of hexagon $BCDEFG$, inscribed in a conic, meet in collinear points. It is enough to prove the theorem when the conic is a circle. Let $CD$ meet $FG$ at $N$. $N$ is the *pole* of the line joining the vertex, $I$, between $C$ and $D$, to the vertex, $L$, between $F$ and $G$. Likewise, $MJ$ is the *polar* of $EF \cap CB$, and $HK$ is the polar of $ED \cap GB$. By Brianchon's Theorem, the three *polars* are concurrent at a point, $X$. Therefore, by the *pole-polar* property, the corresponding *poles* are collinear on the *polar* of $X$.                                                                                                 □

**Fig. 7.8** Converse of
Pascal's Hexagon Theorem

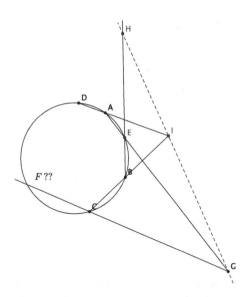

We also have the

**Theorem 7.8 (Converse of Pascal's Hexagon Theorem)** *Let points $A$, $E$, $B$, $C$, $D$ lie on a circle (conic), and suppose there is a point $F$ so the opposite sides of hexagon $AEBCFD$ meet in collinear points. Then $F$ also lies on the circle (conic).*

**Proof** See Fig. 7.8. Let $AD \cap BC = I$ and $EB \cap FD = H$. Draw the line $IH$. Let line $AE$ meet line $IH$ at $G$. Let line $GC$ meet the circle at point $F_1$. Draw line $F_1D$. By Pascal's Hexagon Theorem, the opposite sides of hexagon $AEBCF_1D$ meet in collinear points, on line $GI$. But the side opposite $F_1D$ is $EB$, which meets $GI$ at $H$. So $F_1 = F$.                                                                              □

## 7.5   Notes and Exercises

**Note 1**  The *Circle-to-Circle Theorem* was, essentially, stated by Poncelet as the *Quatrième Principe* and used by Poncelet in his proofs of the Pascal and Brianchon Theorems in *Cahier* 3 of 1813. It was proved in *Cahier* 7, in a three-dimensional argument in which an oblique cone is created whose circular base is the given circle.

Then in Poncelet's proofs of the Pascal and Brianchon Theorems in *Cahier* 3 of 1813, he let the line on $EF \cap BC$ and $FG \cap DC$ in the original circle, pictured in Fig. 7.7 Left, be the vanishing line of a collineation that maps the given circle to one whose center is $X$. So, starting with the diagram of Fig. 7.7, we would have a circle with center $X$ and with $EF \parallel BC$ and $FG \parallel DC$. Then both the Pascal and Brianchon Theorems can be proved. That proof is Exercise 3, below.

**Fig. 7.9** Hexagon Theorem

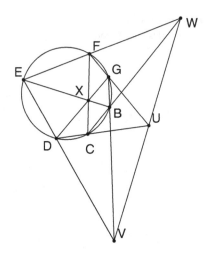

**Exercise 1** See Fig. 7.9. Prove the *Hexagon Theorem*, found in [1, p. 300], without assuming Pascal's Hexagon Theorem: Suppose a hexagon $BCDEFG$ is inscribed in a circle or conic. If the three diagonals on opposite vertices are concurrent, then opposite sides meet in collinear points. (Suggestion: Start with the Circle-to-Circle Theorem, making $X$ the center of a circle.)

Then show by a counterexample that the converse is false.

**Exercise 2** Prove that Pascal's Hexagon Theorem Implies Brianchon's Theorem.
To follow, initially, the train of claims in Brianchon's 1806 proof, we take a hexagon circumscribed about a circle with points of tangency $B, C, D, E, F, G$. By Pascal's Hexagon Theorem, which Brianchon learned from Lazare Carnot, opposite sides of hexagon $BCDEFG$ meet in collinear points $Q, R, N$. See, again, Fig. 7.7 Left. Then find the pole of line $QRN$. (Brianchon had his own proof, based on Desargues' Theorem, of the pole-polar relationship.)

**Exercise 3** Finish the proof of the Pascal and Brianchon Theorems after the original configuration is transformed to that indicated in the Note 1 above, namely, the inscribed hexagon $BCDEFG$ has $EF \parallel BC$ and $FG \parallel DC$.

## 7.6   Some Hints and Solutions

**Exercise 1** See Fig. 7.9. Suppose the opposite vertices are joined by lines—$CF$, $DG$, and $BE$—that are concurrent at $X$. By the Circle-to-Circle Theorem, there is a central collineation whose vanishing line is $x$, the polar of $X$, which maps the configuration to a circle whose center is $X$. Now we have a circle with an inscribed hexagon whose diagonals joining opposite vertices lie on the center of the circle. By considering inscribed angles, we show the opposite sides are parallel. This means that each of the three pairs of opposite sides meet on one line, the line at infinity. So in the original figure, opposite sides meet in collinear points. For the converse, note that we are considering an inscribed hexagon, while Brianchon's Theorem is about a circumscribed hexagon.

**Exercise 2** Solution: See Fig. 7.7 Left. The vertices $I$ and $L$, of the circumscribed hexagon, are on the polar of point $N = CD \cap FG$, the point at which opposite sides meet. We reach the corresponding conclusion for the polars of $Q$ and $R$. Then by the *pole-polar* property, the polars of those three collinear points are concurrent. And that is what we need to show.

# References

1. Charles Julien Brianchon, Sur les surfaces courbes du second degré, *Journal de l'École Polytechnique* Cahier 13, Tome 6, 1806, 297–311.
2. Lazare Carnot, *Géométrie de Position*, Paris: Duprat, An XI, 1803.
3. J. V. Field and J. J. Gray, *The Geometrical Work of Girard Desargues*, New York: Springer, 1987.
4. Philippe de La Hire, *Nouvelle Méthode en Géométrie pour les Sections des Superficies coniques et Cylindriques*, Paris, 1673.
5. Philippe de La Hire, *Sectiones Conicae en novem libros distributae*, Paris 1685; French translation by Jean Peyroux, *Grand Livre des Sections Coniques*, Paris: Blanchard, 1995.
6. Blaise Pascal, *Essay pour les coniques*, 1640, in [9] 190–194, in English transl. in [3], 180–184.
7. J V Poncelet, *Applications d'analyse et de géométrie qui ont servi, en 1822, de principal fondement au traité des propriétés projectives des figures, etc.*, 2 tomes, Paris: Mallet-Bachelier, 1862–64.
8. J. V. Poncelet, *Traité des Propriétés Projectives des Figures*, Paris: Bachelier 1822.
9. René Taton, *L'oeuvre mathématique de G. Desargues*, Paris: Presses Universitaire de France, 1951.

# Chapter 8
# Nineteenth Century

## 8.1   Monge and Carnot: Steps Toward Projective Geometry

Gaspard Monge (1746–1818) led a revival of geometry in France. Despite his background as a commoner, his talent brought him in 1765 to the school of military engineering at Mézières, then the leading school in France for scientific studies. He became a professor of mathematics in 1769. At that time he wrote groundbreaking papers in what we now call the differential geometry of surfaces and he developed *descriptive geometry*. That study, installed in the curriculum at Mézières, was to train future engineers in the practical mathematics expected in the architecture of fortifications and in the theory that underlay the mathematics. In 1784 Monge left the school at Mézières to devote himself to several positions of public service, including examiner of naval officer students, and later as Minister of the Navy. He supported the French Revolution; in 1793 he organized the manufacture of the iron and steel and armaments and gun powder needed for French military victories that saved the new French Republic in war with its neighbors. Soon after, he became a principal founder of the short lived *École Normale* and of the *École Polytechnique* [20]. With his influence, descriptive geometry took an important place in the curriculum of the *École Polytechnique*.

Among Monge's students at Mézières was Lazare Carnot (1753–1823). In the French Revolution, he became a member of the 12-person Committee of Public Safety in the period called the Terror, and he was responsible for building the army by conscription and organizing that army to successfully defend France in 1793. Less widely known, Carnot was a mathematician. He introduced several projective concepts in his *Géométrie de Position* of 1803. Whereas La Hire had, after a brief definition, looked at harmonic sets as formed by a line cutting a circle, involving tangents to the circle, on the road to the study of conic sections, Carnot began with a *complete quadrilateral*. He showed that when opposite sides of complete quadrilateral $ABCD$ meet at $X = AB \cap CD$ and $Y = BC \cap AD$, and diagonals $AC$ and $BD$ cut line $XY$ at, respectively, $W$ and $Z$, then $H(XY, ZW)$. In other

© Springer Nature Switzerland AG 2020
C. Baltus, *Collineations and Conic Sections*,
https://doi.org/10.1007/978-3-030-46287-1_8

**Fig. 8.1** Complete
quadrilateral $ABCD$,
$H(XY, ZW)$

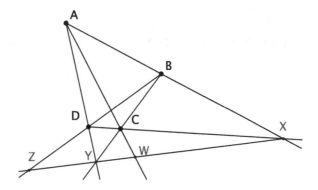

words, $X$ and $Y$ are harmonic conjugates of $Z$ and $W$, or, as Carnot phrased it, $XY$ is cut in *segmens proportionnels*. See Fig. 8.1. This was Theorem VIII of [5, p 282] of 1803. Application to circles and then to the conic sections came only afterwards. He also introduced signed segment lengths. So, no matter what the order of collinear points $A, B, C$, $AB + BC = AC$. And the order of points needed no separate consideration in defining harmonic conjugates; $H(AB, CD)$ exactly when $AC \cdot BD = -(AD \cdot BC)$, when the lengths are signed. We have already seen what was yet another contribution of Carnot, expressing the ratio defining a harmonic set $A, B, C, D$ in terms of the sines of angles at a point $O$, not on line $AB$, formed by rays $OA, OB, OC, OD$ [6, p. 93].

## 8.2   Jean-Victor Poncelet

This brings us to Jean-Victor Poncelet (1788–1867). He entered the *École Poly-technique* in November, 1807. Under the influence of Gaspar Monge, the *École Polytechnique* at the time was alive with geometry. Students produced papers on geometry, to appear in the *Journal de l'École polytechnique* and the more informal *Correspondance sur l'École polytechnique*. Michel Chasles (1793–1880), himself a graduate of the *École Polytechnique*, wrote in an address of 1846,

> I have spoken at length on the works of Monge and Carnot because I regard them as having reanimated in France the spirit of geometric methods and having inspired the young mathematicians who, soon after, entered this field [7, p lxxx].

Among those young mathematicians then mentioned by Chasles was Poncelet.

Poncelet is the person most often credited with the creation of modern projective geometry [8, p 92], [9, p ix]. John Wesley Young wrote in 1930 [21, p 9]

> projective geometry as a self-contained discipline was not developed until the great French mathematician Poncelet (1788–1867) published his classic *Traité des propriétés projectives des figures* in 1822.

This is a reasonable judgment. Where, before Poncelet, projective geometry constituted a method or a chapter in the study of the conic sections, Poncelet shifted attention to the projective geometry itself. Lazare Carnot had already started on this path. Poncelet's work was about projection, the projection of a primitive plane figure onto a figure in the same plane or another plane, and about *invariants*—not Poncelet's terminology, properties unchanged by projection. Later in the nineteenth century, Felix Klein's *Erlanger Program* organized distinct geometries by their invariants.

We'll look at Poncelet's solution, as a student, of the Problem of Apollonius. Augustin-Louis Cauchy (1789–1857), who had entered the *École Polytechnique* in 1805 at age 16, had produced a solution in 1806 [4]. This is the problem of constructing a circle tangent to three given circles. Apollonius was reported to have a solution, and modern solutions go back to Viete. Lazare Carnot's widely read *Géométrie de Position* of 1803 [5] gave a solution involving coordinates. Poncelet as a student gave an elegant solution in synthetic geometry and would later return to the problem—a sort of test case in geometry—several times.

Poncelet wrote three comprehensive works on projective geometry. The first was a set of seven *Cahiers* or *Notebooks* (1813) written while Poncelet was a military prisoner in Saratov, on the Volga, in the 15 months after his capture during Napoleon's wintry retreat from Moscow, in November, 1812. Poncelet was left without books, but it is clear he had imbibed deeply the geometry of the *École Polytechnique*. The second was *Essai sur les propriétés projectives des sections coniques*, presented to the *Académie des Sciences* in 1820. Poncelet had these two works published in the 1860s [17, 18].

The third is his masterwork, *Traité des propriétés projectives des figures*, published in 1822. It incorporated and developed the earlier work; in the case of central collineations, it includes steps well beyond what had appeared earlier.

We first introduce the particular central collineation that is now called *dilation*, and, later, the one that Poncelet referred to as an *inverse homology*.

## 8.3 Dilations and the Inverse Homologue

**Definition 14** A *Dilation* (in $\mathbf{R}^2$) has a *center*, a point, $O$, and a non-zero numerical *scale factor*, $k$. In the coordinate plane, the dilation with center $(0, 0)$ and scale factor $k$ maps each point $(x, y)$ to $(kx, ky)$. In more general terms, any point $X$ is mapped to the point $X'$ on line $OX$ so that, for distances, $|k| \cdot OX = OX'$, with $O$ between $X$ and $X'$ exactly when $k$ is negative.

We now say that two figures are *similar* if one is congruent to an image of the other under a dilation. Poncelet said as much in [19, 1822 Art. 241].

The *Centers of Similitude* of a pair of circles are the centers of the two dilations which map one circle to the other. (Concentric circles have only one Center of Similitude.) The points at which common tangents meet are centers of similitude.

**Fig. 8.2** $S_1$ and $S_2$ are centers of similitude

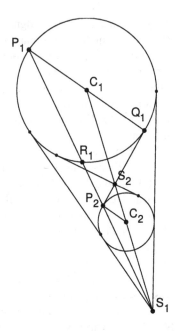

The centers are $S_1$ and $S_2$ in Fig. 8.2. In general, if parallel radii of circles $C_1$ and $C_2$ meet their respective circles at $P_1$ and $P_2$, then the two points at which the collection of lines $P_1 P_2$ meet are the two *Centers of Similitude*, $S_1$ and $S_2$. When the dilation has a positive scale factor, the related parts of the diagram are *direct homologues* and when the scale factor is negative, the related parts are *opposite homologues*. In Fig. 8.2, $P_1$ and $P_2$ are direct homologues, while $Q_1$ and $P_2$ are opposite homologues. When two points, one on each circle, are collinear with the center but not images of each other under the dilation, they are called *inverse homologues*. In Fig. 8.2, $P_2$ and $R_1$ are inverse homologues, with center $S_1$.

Poncelet, and many future writers, used the term *homology* for a dilation, and points or parts of diagrams that are related by a dilation are called *homologues* or *homologous*. (The term *homology* later assumed a broader meaning.)

Leonhard Euler (1707–1783) had referred to *similar* figures in Book 2 of [11, 1748], Art. 439, and used the term *homologous*. He later introduced the concept in *De centro similitudinis* [12], but the dilation was just a particular case of the mapping Euler called a *similitude*. The concept of dilation was rarely employed before Poncelet, who made it a core of his work. Poncelet called a dilation a *similitude*, and called figures related by a dilation *s. and s. p.*, *similar and similarly placed*.

In Theorem 8.1, we examine the dilation as a central collineation in $\mathbf{P}^2$. What is the line of fixed points? (We'll see the answer below; Poncelet answered in Art 329. of 1822.) We will later show that mapping points to their inverse homologues is effected by a central collineation.

**Fig. 8.3** Theorem: dilation is
a central collineation

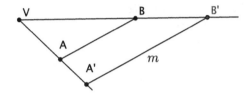

**Theorem 8.1 (Dilation as Central Collineation)** *Given distinct collinear points
V, A, and A′, the (V, l) collineation mapping A to A′, where l is the line at infinity,
is the dilation with center V and scale factor k = VA′/VA where the lengths are
signed.*
*Further, all lengths are multiplied by |k| and a line and its image are parallel.*
*The image of a circle, with center $C_1$ and radius $r_1$, under a dilation with scale
factor k, is a circle whose center, $C_2$, is the image of $C_1$ and radius is $r_2 = |k| \cdot r_1$.*
*A point where two internal or two external common tangents meet is a center of a
dilation.*

**Proof** See Fig. 8.3. Take a point B not on VA. We'll show that B′ found by the
given central collineation is the image of B found by the given dilation. According
to the collineation, we draw AB and a line m on A′ parallel to AB since the axis is
the line at infinity. B′ is, then, m ∩ VB. However, by parallels and the Side Splitter
Theorem, $\dfrac{VA'}{VA} = \dfrac{VB'}{VB}$. So $\dfrac{VB'}{VB} = k$. This means the image of B is the same under
the central collineation as under the dilation. We can now see the same result in the
case where B is on line VA.

By the same diagram, using the Side Splitter Theorem and triangle similarity, we
see that a dilation multiplies all lengths by |k| and maps lines to parallel lines.

When mapping a circle with center $C_1$, since a dilation multiplies all lengths by |k|,
the points at distance r from $C_1$ are mapped to the points at distance |k|r from the
image of $C_1$, so the image of a circle is a circle.

Any tangent to circle $C_1$ is mapped by a dilation to a parallel tangent to circle $C_2$ so
the point at which common tangents meet is the center of that dilation.          □

We turn to Poncelet's student paper, of 1809, which appeared in *Correspondance
sur l'École polytechnique* [15]. In it he presented a simple and elegant solution to
the Problem of Apollonius. It may have been an entry by Poncelet into projective
geometry.

The mathematical background called on in [15] is, in several propositions,
developed and proved at the beginning of *Cahier* 1 of 1813, and we take advantage
of the 1813 work in setting out the basics for understanding the 1809 solution.

We remind the reader of several theorems, where 1. and 3. are in Chap. 3, and
then provide further background as found in *Cahier* 1, of 1813.

1. *Power-of-a-Point Theorem*: In Fig. 8.4, $AP \cdot EP = PD^2$ when PD is tangent
   to the circle, and AEP is any secant on P.

**Fig. 8.4** Power of a Point
Theorem, Point P

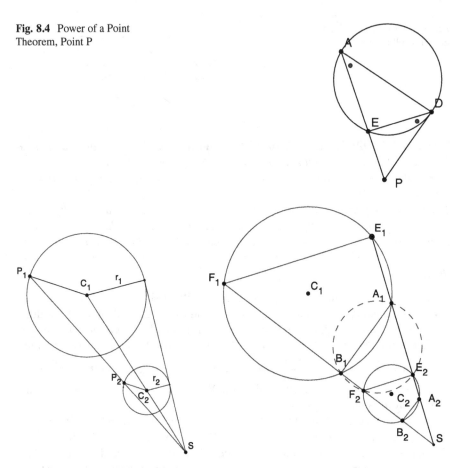

**Fig. 8.5** Left: Poncelet 1813 Prop. 1      Right: Poncelet 1813 Prop. 3

2. *Cyclic Quadrilateral Theorem.* A quadrilateral is *cyclic*, i.e., can be inscribed in
   a circle, if and only if opposite angles are supplementary.
3. *Common Chord Theorem.* If three circles meet pairwise, then the common
   secants or chords are concurrent.

**Poncelet 1813,** *Cahier* **1 Prop. 1** (See Fig. 8.5 Left). Let $C_1$ and $C_2$ be centers
of circles, with center of similitude $S$, and with parallel radii $C_1 P_1$ and $C_2 P_2$ as
pictured. Then by similar triangles $\dfrac{r_1}{r_2} = \dfrac{C_1 S}{C_2 S} = \dfrac{P_1 S}{P_2 S}$, and $P_1$, $P_2$, and $S$ are
collinear.                                                                                    □

**Poncelet 1813,** *Cahier* **1 Prop. 3** (See Fig. 8.5 Right. Following pairs of
supplementary angles proves *a.*, below.) Let $C_1$ and $C_2$ be centers of circles, with
center of similitude $S$, and inscribed quadrilaterals $A_1 B_1 F_1 E_1$ and $A_2 B_2 F_2 E_2$
corresponding under a dilation with center $S$. Then

*a*. Quadrilaterals $A_1 B_1 F_2 E_2$ and $A_2 B_2 F_1 E_1$ are cyclic quadrilaterals, and

*b*. $SA_1 \cdot SE_2 = SB_1 \cdot SF_2$, where these products are the same no matter how we select inscribed quadrilateral $A_1 B_1 F_1 E_1$.

For *b*., note that

$$\frac{SE_1}{SE_2} = \frac{r_1}{r_2} = \frac{SF_1}{SF_2},$$

and $SE_1 \cdot SA_1 = SF_1 \cdot SB_1$ by the power-of-point $S$ with respect to circle $C_1$. So

$$(SE_2 \cdot \frac{r_1}{r_2})SA_1 = (SF_2 \cdot \frac{r_1}{r_2})SB_1.$$

Then simplify.    □

**Poncelet 1813, Corollary to Prop. 3** In Fig. 8.5 Right, let $A_1$ and $B_1$ coalesce. Then the circle on $A_1$, $B_1$, $F_2$, and $E_2$ becomes tangent to the two given circles at points we call $T_1$ and $T_2$, with $S$, $T_1$, and $T_2$ collinear.

We proceed to the 1809 solution of the Problem of Apollonius. Poncelet, as had Viete, first reduced the radii of the three given circles by the least radius, $r_3$, transforming the problem into that of constructing a circle on a given point, $A$, and tangent to two circles, $C_1$ and $C_2$. One then reduces the radius of the constructed circle by $r_3$.

Take, as pictured as pictured in Fig.8.6, inverse homologues, $E$ on circle $C_1$ and $D$ on circle $C_2$.

**Claim** There is a point $B$ on line $SA$, independent of the choice of $E$ and $D$, that lies on all of the circles $ADE$.

**Proof of the Claim** By Poncelet's *Prop*. 3, the product $SE \cdot SD$ is the same for all choices of $D$ and $E$. But $SE \cdot SD$ is the power of $S$ with respect to circle $ADE$, and must equal $SA \cdot SB$ when $B$ is the second point at which line $SA$ meets circle $ADE$. So $B$ is determined (Fig. 8.6).

Now consider the three circles: $C_1$, circle $ADE$, and the circle (to be found) on $A$ and $B$ and tangent to circle $C_1$. The three common secants are concurrent at a point $P$. $P$ lies on $AB$ and on the common secant of circles $C_1$ and $ADE$. So the point at which a circle on $A$ and $B$ is tangent to circle $C_1$ is a point of tangency $T_1$ to circle $C_1$ from $P$, and this tangent can be directly constructed. The circle sought is that on $T_1$, $A$, and $B$.    □

## 8.4 The Ideal Common Secant and Homology, 1813

Poncelet was given to pondering the fate of mathematical objects which disappear with a continuous variation in the figures or equations that define those objects. He

**Fig. 8.6** Solution, problem
of Apollonius, Poncelet
1809/1813

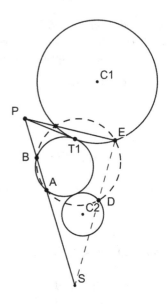

saw those mathematical objects becoming *ideal* or *imaginary*. The most important
of these was the common secant of two circles, already applied in 1809. Now, what
happens to the common secant when the circles separate? Poncelet took this up
in *Cahier* 1 of 1813. (In a case of simultaneous discovery, while Poncelet was a
prisoner in Russia, a paper by a fellow graduate of the *École Polytechnique* appeared
on the same topic, conferring the name *radical axis* to Poncelet's *common secant*
[13].)

A modern student may already have encountered the common secant of two
circles. Suppose you want to find the points of intersection of two circles, say
$x^2 + y^2 + 4x = 0$ and $x^2 + y^2 + 2x - 4y = 4$. When you subtract the two equations,
you get $2x + 4y = -4$ or $x + 2y = -2$. That line, with equation $x + 2y = -2$, is the
line on the two points at which the circles meet, the common secant of the circles.
But what happens if the circles do not meet and you carry out the corresponding
computation? You still get a line, still called the common secant, or *radical axis*, of
the circles, but when you compute its intersection with the circles, you get imaginary
values. The circles meet at *imaginary* points, but the common secant is real and
*ideal*, as Poncelet labeled it.

In 1813, Poncelet settled on a property of tangent length to characterize the
common secant of two circles, whether they meet or not. (He used a different
property in [18].) The property is set out in *Cahier* 1 Prop. 5 and 6, followed by
a definition.

**Poncelet 1813,** *Cahier* **1 Prop. 5** (See Fig. 8.7 Left.) Let $S$ be a center of
similitude for circles $C_1$ and $C_2$, and let $T$ and $t$ be inverse homologues, respectively.
Let tangents at $T$ and $t$ meet in $\alpha$. Then tangents $\alpha T$ and $\alpha t$ are equal.

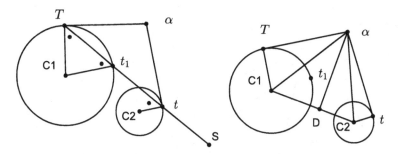

**Fig. 8.7**   Left: Poncelet 1813 Prop. 5        Right: Poncelet 1813 Prop. 6

*Proof* Radii to $t$ and its homologue $t_1$ are parallel and so form equal angles with $ST$. With equal angles at $T$ and $t_1$, then angles $Tt\alpha$ and $tT\alpha$ are equal, so $T\alpha = t\alpha$.
□

**Poncelet 1813,** *Cahier* **1 Prop. 6** (See Fig. 8.7 Right.) The set of all such points $\alpha$ forms a line perpendicular to $C_1C_2$, the *common secant* of the two circles. The proof is part of Theorem 8.2., below.

**Definition 15** For two given circles, if the circles meet in two points, then the line on those two points is their *Common Secant*, also called the *Radical Axis*. When the circles do not meet in two points, then the common secant is the set of points whose tangents to the circles are equal in length. In other words, these are the points whose *powers* to the two circles are equal.

Now we consider inverse homology as a mapping.

**Theorem 8.2 (Inverse Homology as Central Collineation)** *Given circles $C_1$ and $C_2$ with Center of Similitude $S$, take the dilation mapping circle $C_1$ to $C_2$, and point $E$ on $C_1$ to $E'$ on circle $C_2$. Let $E''$ be the other point of $C_2$ on line $SE$. (So $E''$ is the inverse homologue of $E$.) Then the $(S, l)$ collineation mapping $E$ to $E''$ maps circle $C_1$ to circle $C_2$, where $l$ is the common secant of the two circles.*

*Proof* See Fig. 8.8. Let $\phi_1$ be the dilation with center $S$ that maps point $C_1$ to $C_2$. Let $\phi_2$ be the involution collineation with center $S$ that maps point $E'$ to $E''$ (and, so, maps circle $C_2$ to itself). The composition $\phi_2 \circ \phi_1$ is a collineation with center $S$ so it has a line of fixed points, $l$. And it maps each $E$ to $E''$ as indicated above. The tangent on any point $E$ of circle $C_1$ would be mapped by a central collineation to the tangent at $E''$ to circle $C_2$. Thus those tangents meet on the axis. The tangents at $E$ and $E''$ form equal angles with line $ES$, so the axis, $l$, is the common secant. Since circles $C_1$ and $C_2$ are symmetric about the line on circle centers $C_1$ and $C_2$, then $l \perp C_1C_2$.
□

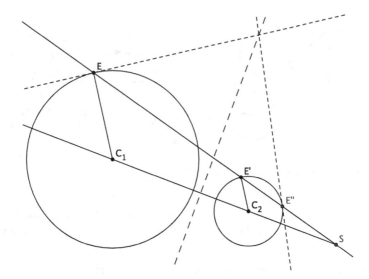

**Fig. 8.8**  Theorem 8.2

**Fig. 8.9**  Based on Fig. 14
from Poncelet's *Cahier* 1,
1813/1862

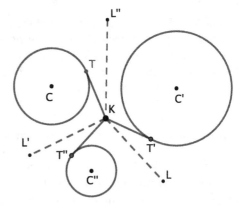

We observe that parallel tangents at corresponding points of the two circles are
equal in the sense that both are infinitely long. In this way, the line at infinity is a
second common secant, an idea that Poncelet takes up in his 1822 work.

With this formulation of the common secant, Poncelet extended the Common
Secant Theorem to the case where the three circles do not meet. This case is
illustrated in a drawing by Poncelet from Poncelet's 1813 *Cahier* 1 Prop. 8, in
Fig. 8.9, redrawn 1862. From point $K$, tangent segments to the three circles are
equal.

As we shall see, the common secant of two circles will evolve into the line of
fixed points, or *axis*, for two homologous figures.

## 8.5   More Material in Poncelet's *Cahiers* of 1813–1814

After the initial material noted above, Poncelet completed *Cahier* 1 with construction problems, including a more elaborate solution of the Problem of Apollonius, which displayed the power of the notion of the common secant. *Cahier* 2 concerns coordinate geometry. *Cahier* 3 begins with a claim that his method, at once both geometric and analytic, to demonstrate properties of figures, will arrive at properties that are impossible or very difficult to show by either geometric or analytic methods alone. The method, which Lazare Carnot and Charles Julien Brianchon, especially, had employed, transforms a problem to a simpler one.

The method was a projection in space, whereby a conic was projected to a circle, a line was projected to the line at infinity, or both. This concept had been central to Desargues and to La Hire, but, in 1813, Desargues' work had been lost and La Hire little studied. Among the principles noted were those of a plane-to-plane projection in space. As an example, a figure with several sets of concurrent lines whose points of concurrency are collinear, can be projected to a plane so the concurrent lines become systems of parallel lines. The line on which the points of concurrency lie is projected to the line at infinity. Brianchon, another former student of the *École Polytechnique*, wrote "I carry out 'la perspective' onto a plane parallel to that determined by the eye and the [given] line of [fixed points] ... so that the sides, in place of turning on the fixed points move in a parallel fashion among themselves." This quote is from [3, p 2], the one work that Poncelet explicitly noted in his book of 1822 as a providing "the first idea of this work" [19, p xxiv]. Poncelet applied such projections in several proofs, including that of Pascal's Hexagon Theorem and its dual theorem, which had been proved by Brianchon in [2].

Poncelet applied his methods in *Cahier* 4 to problems of inscribing and circumscribing polygons to conics, problems which call to mind those in Carnot's [5].

These projections were justified by Poncelet's *Quatrième Principe*, [17, p 381–386] at the beginning of *Cahier* 3, namely, that given any circle and outside line, there is a projection with the given line as the vanishing line and by which the circle is mapped to another circle. We gave a central collineation with this property as the *Circle-to-Circle Theorem*, Theorem 7.5, but Poncelet worked in three-dimensions, showing how to create an oblique cone and a particular plane slicing the cone so projection to that slicing plane gives the projection we need. The slicing plane is parallel to the plane on the vertex of the cone and on the vanishing line, and slices the plane in the *subcontrary* circle of the cone, as described in Proposition 5 of Book 1 of Apollonius. (Oblique cone: The axis of the cone is not perpendicular to the base plane.)

In *Cahier* 3, Poncelet had noted the *pole-polar* property of conic sections, referring to Monge's geometric development of the idea in three-dimensional space in [14, Art. 44].

We have a proof in *Cahier* 7, p 439, of the *pole-polar* property by application of Poncelet's *Quatrieme Principe*. See Fig. 8.10. On the left we have the given conic and outside line $LM$. For points $m, m', m''$ on $LM$, the polars are constructed by

**Fig. 8.10** Based on
Poncelet's Fig. 175 in *Cahier*
7, proof of the Pole-Polar
property, 1813/1862

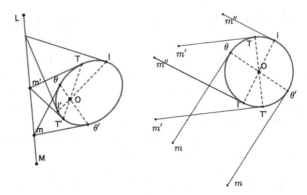

drawing tangents to the conic and then, for each of $m$, $m'$, $m''$, joining the points of
tangency. To prove that those polars are concurrent, the conic is mapped to a circle
and line $LM$ to the line at infinity, giving the diagram on the right. Since the pairs
of tangents are mapped to parallel tangents to the circle, the polars are mapped to
diameters. Diameters of a circle are concurrent, so the polars are concurrent (at the
pole of $LM$.).

The dual, which Poncelet called *le réciproque* in 1814, followed immediately:
concurrent lines on a point $O$ have collinear poles (on the polar of $O$).

At this point Poncelet ended his *Cahiers*, "Brusquely interrupted at Saratoff, in
June 1814, at the notification of the general peace" [17, p 441].

## 8.6   Poncelet's *Traité* of 1822

After his return from Russia in the summer of 1814, Poncelet became a captain
of the engineering corps at Metz. He read, reflected, and wrote on projective
geometry, producing several articles for Gergonne's *Annales de mathématiques
pures et appliquées*. In 1820 he presented [18] to the Academy of Sciences. It was
a refinement of the *Cahiers* of 1813–14.

Then, in just a few years, Poncelet reworked and expanded his paper of 1820
into a deeper work, published in 1822 as *Traité des propriétés projectives des
figures*. He tells us in the introduction that his goal is to perfect and generalize
this geometry and to "render it independent of algebraic analysis" [19, p. xix]. He
reiterated the purpose, which he had stated in *Cahier* 7, p 373, of his 1813–1814
study: "to demonstrate . . . the class of properties of figures whose parts fit together
under general conditions of position, independent of all size and determinate
measure." Where, with Monge, projections had been parallel projections in space,
now Poncelet studied plane-to-plane projection from a point not on either plane,
where "all that is projected runs toward a point or *center* of projection" [19, p. xxxii].
When we rotate one plane to lie on the other, we recognize a central collineation—
not Poncelet's term—and a conic section is the projection of a circle.

**Fig. 8.11** Left: based on
Carnot's Fig. 8, 1806.
Right: based on Poncelet's
Art. 39, 1822

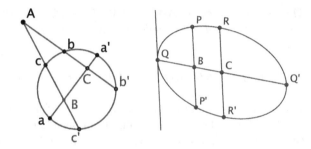

This can be seen as a step into modern geometry: instead of jumping into properties of figures, Poncelet started by introducing a transformation and telling us that his subject encompassed those properties preserved under the transformation.

In Chap. 1, Poncelet took up the question of which properties would be preserved under projection. The first is that known to the ancient Greeks, a line cut in harmonic proportion: four collinear points $A, B, C, D$ where $CA/CB = DA/DB$ (and exactly one of $C$ and $D$ separates $A$ and $B$). Further, following La Hire (1685), four concurrent lines lying on points in harmonic proportion form a *harmonic pencil* [p. 13]. (Poncelet used the term "faisceau"; we will employ the modern "pencil.") Referring to Carnot's proof, Poncelet noted that four points in harmonic proportion are projected to points in harmonic proportion. And we have points in harmonic proportion when one is at infinity and a second is the midpoint of the other two. Further, a pencil of lines will concur at infinity when the lines are parallel [p. 15]. As we showed in examining the work of La Hire, when $H(AB, CD)$ and $O$ is the midpoint of $CD$, then $DO^2 = CO^2 = OA \cdot OB$ [p. 18].

An important advance in Chap. 1 was the extension of the invariance of the harmonic relation under projection to other particular sets of collinear points. We recall in Carnot's proof of Theorem 5.4, invariance of the harmonic relation under projection, the argument that when collinear points $A, B, C, D$ are joined to outside point $E$, then

$$\frac{AD \cdot BC}{AB \cdot CD} = \frac{sin(\angle AED)}{sin(\angle AEB)}\frac{sin(\angle BEC)}{sin(\angle CED)}.$$

This thinking was extended in Poncelet's Art. 20 to other products of ratios, including, as we will see in the work of Jacob Steiner, the cross-ratio, $\dfrac{AD \cdot BC}{AB \cdot CD}$. In summary:

**Theorem 8.3 (Poncelet's Art. 20, 1822: Carnot's Principle)** *Suppose we have a ratio involving collinear points that can be expressed as the product of ratios of the sines of angles from an outside point $E$ drawn to those collinear points. As with Carnot's argument, that product of ratios will be invariant under projection from $E$.*

Poncelet applied this principle, in Art. 34, to Carnot's theorem from 1806 [6] Art. 9.

**Theorem 8.4 (Poncelet Art. 34 of 1822, based on Carnot 1806)** *When triangle ABC meets the sides of a conic in R, R' on AC, in Q', Q on BC, and in P', P on AB, then*

$$AP \cdot AP' \cdot BQ \cdot BQ' \cdot CR \cdot CR' = AR \cdot AR' \cdot BP \cdot BP' \cdot CQ \cdot CQ'.$$

***Proof*** Carnot proved the theorem in the case of a circle in Article 9 of [6]. See Fig. 8.11 Left, which illustrates that from outside point $A$, $Ac \cdot Ac' = Ab \cdot Ab'$ (Power-of-a-Point Theorem). We get a similar equation for point $B$, and again for point $C$. Then multiplication gives the claimed equation in the case of a circle. This equation can be regarded as a ratio equal to 1. We then apply Art. 20 to see that, for example, $\dfrac{CR \cdot CR'}{AR \cdot AR'}$ is unchanged under projection. So the equation for the circle holds for a conic.                                                                                    □

Then we have important applications of this formula. This example is Poncelet's Art. 35. In the formula of Art. 34, we project so $A$ is mapped to infinity. Therefore $BP \parallel CR$. With $A$ at infinity, $\dfrac{AP \cdot AP'}{AR \cdot AR'} = 1$. Therefore, by Art. 34,

$$\frac{BP \cdot BP'}{BQ \cdot BQ'} = \frac{CR \cdot CR'}{CQ \cdot CQ'}. \tag{8.1}$$

In Art. 39, we let line $QQ'$ be the diameter with midpoint $C$, with $RR'$ the corresponding ordinate on $C$, and with $PP'$ a parallel ordinate. See Fig. 8.11 Right. Since $(CR \cdot CR')/(CQ \cdot CQ')$ is a constant, $p$, then no matter how ordinate $BB'$ is selected (with midpoint $P$, so $BP = BP'$), equation (8.1) becomes the abscissa-ordinate equation of the conic:

$$\frac{BP^2}{BQ \cdot BQ'} = p \quad \text{or} \quad BP^2 = p \cdot BQ \cdot BQ'.$$

Poncelet derived other properties. Speaking of the collection of properties, Poncelet wrote in Art. 39,

> From this one would immediately deduce all the definitions and known properties of the *center, axes, conjugate diameters*, and *asymptotes* of conic sections.

This 1822 work is filled with conic properties that are proved by a projection. Here is just one more, in Art. 186. Our Fig. 8.12 Left shows a conic with an inscribed quadrilateral, $ABCD$, whose opposite sides meet at $E$ and $F$, and the circumscribed quadrilateral whose sides are the tangents at $A$, $B$, $C$, $D$. Our Fig. 8.12 Right, based on Poncelet's *Fig.* 27, shows the figure after the conic is projected, with vanishing line $EF$, so the conic becomes a circle. Since both pairs of opposite sides are parallel, then in a circle, $ABCD$ must now be a rectangle. Further, the tangents

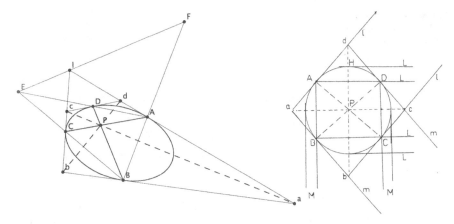

**Fig. 8.12** Left: Conic, based on Poncelet's Art. 186, 1822
Right: Based on Poncelet's *Fig.* 27, 1822

drawn on the four vertices must form a rhombus, so its diagonals meet at the center of the circle, $P$. Conclude about the original conic:

1. The diagonals of the inscribed and circumscribed quadrilaterals are concurrent;
2. $P$, where the diagonals meet, is the pole of the line $EF$ (providing an easy construction of the polar of a point inside a conic), and the points at which opposite sides of the circumscribed quadrilateral meet lie on $EF$;
3. The diagonals of the circumscribed quadrilateral lie on $E$ and $F$.

## 8.7   Poncelet in 1822: Inverse Homologues, the Common Secant as Axis and Vanishing Line

In his 1822 *Traité*, Poncelet brought together several concepts introduced in his *Cahiers* of 1813–1814. First, two circles, $c_1$ and $c_2$, are related by a *direct similitude*, a dilation, with center $S_1$, and by a second dilation, the *opposite similitude* with center $S_2$ [Art. 237]. The *inverse homologue*, $X'$, of a point $X$ on circle $c_1$ is the point on circle $c_2$ which lies on ray $OX$ and where the tangent at $X'$ meets the tangent at $X$ (at a finite point). ($O$ can be $S_1$ or $S_2$.) And, as we have seen, as $X$ ranges along circle $c_1$, the tangents at $X$ and $X'$ meet along a line, called the *common secant/chord* of circles $c_1$ and $c_2$. Further, when a point $X$ of $c_1$ is related to a point $X''$ of $c_2$ by the direct or opposite similitude, the tangents at $X$ and $X''$ are parallel; by analogy, we say those corresponding tangents meet on a second *common secant*, the line at infinity.

Poncelet gave a visual way of relating the two common secants for two circles. Imagine the circles lie in distinct but parallel planes. We have parallel tangents to the two circles meeting on the line at infinity where the parallel planes meet. Now allow one plane to rotate with respect to the other until the common secant at infinity

**Fig. 8.13** Subcontrary
circles on planes meeting in a
common secant

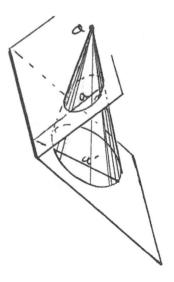

becomes the finite common secant (Fig. 8.13). Fixing in position the two planes
after such a rotation, the lines in 3-dimensional space joining corresponding points
of the two circles form a cone and one circle will be the *subcontrary* to the other, as
described in Apollonius Book 1 Prop 5.

Then we forget the cone and rotate one of the planes to lie on the other, so the
two circles lie in this new single plane.

> ...for one can put the figure in projection, on a new plane, so that the proposed circles
> remain circles for which the common secant at infinity becomes the common secant at
> finite distance ... (called *axis of projection* in Article 246)
> ...the same reasonings and the same consequences are applicable to any conic sections
> which are *s. and s. p.* (Article 244)

From this, Poncelet concluded that the properties enjoyed by the common secant
at infinity are also enjoyed by the finite common secant. So since lines which are
direct homologues meet on the common secant at infinity, then inverse homologues
meet on the finite common secant, the radical axis.

As we shall see, these concepts are extended to the conics which rise from
projection of the pair of circles we have considered; as lines are preserved by
projection, there are still *common secants* for conics related this way. And when
the conic arises from a plane slicing a conic surface, a *common secant* of the base
circle and the conic will be the line at which the slicing plane meets the base plane
(Art. 65). This corresponds to La Hire's identification of the line on which a plane
slicing a cone meets the base plane as the axis of a central collineation. (In Art. 269,
Poncelet returned to the problem of a circle tangent to three given circles.)

Further, Poncelet observed that some properties of figures *s. and s. p.* can still
hold for figures which are not *s. and s. p.*. He is moving toward (discovering?)
the central collineation involved, where the center of the central collineation is the
"point at which common tangents meet." From Article 297,

The properties of two conics in a plane, with respect to the point at which common tangents meet, are not particular to those curves; they occur when two figures can be regarded as the projections of two figures similar in size and position. The center of similitude is represented by a point which enjoys, with respect to these corresponding figures, all the properties of a point of intersection of two common tangents.

Thus, two circles in one plane could be projected to a second plane as, say, an ellipse and a hyperbola, no longer *s. and s. p.*, but still *homologous*. The center of similitude is mapped to a point which is collinear with pairs of homologous points: "...one can call it the center of homology of the two figures, ..." and "the line on which the homologous lines meet will be the axis of homology of the system" (Article 298), the axis of a central collineation.

Although Poncelet did not designate this description of homologous figures as a "definition," it is the definition adopted by Chasles in his *Traité de géométrie supérieure* [7, Art. 518]. And it gives precisely the figures related under a central collineation.

Articles 293–294 suggest a second extension of *homologous* to figures not *s. and s. p.*: two figures $K_1$ and $K_2$ in plane $\sigma$ are homologous if they are both projections, from different centers in space, $S_1$ and $S_2$, of one figure on a plane distinct from $\sigma$. The center of homology is the point at which line $S_1 S_2$ meets $\sigma$. This is the definition of "homology" used by Cremona in his 1893 book. Again, two figures $K_1$ and $K_2$ are related this way exactly when they are related by a central collineation [10, p 9].

That straightedge constructions follow was noted in Poncelet's Article 302.

It results from the purely descriptive properties of homologous figures that if a point of one with its homologue is given, and the center and axis of homology, one could entirely trace the figure by means of what is given, using only the straight line. For, as homologous lines must meet on the axis, it all consists of drawing, on the given point, a sequence of transversal lines whose homologues will be exactly determined. ...

Poncelet further noted in Art 304, if one is given two lines and a point and their homologues, then the center and axis can be found. Several similar cases were noted.

Poncelet did not pursue the topic any further, and there is no diagram to illustrate Article 302. Could it be a last-minute realization of an implication of his ideas?

## 8.8 Notes and Exercises

**Note 1** A helpful look at Poncelet's early work is [1].

**Exercise 1** Prove Poncelet's Proposition 5 of his 1813 *Cahier* 1 in the case that the center of similitude lies between the two given circles.

# References

1. Bruno Belhoste, De l'École polytechnique à Saratoff, les premiers travaux géométriques de Poncelet, *Société des amis de la bibliothéque de l'École Polytechnique (SABIX)*, www.sabix. org/bulletin/b19/belhoste.html. (original Bulletin No. 19, 1998).

2. Charles Julien Brianchon, Sur les surfaces courbes du second degré, *Journal de l'École Polytechnique* Cahier 13, Tome 6, 1806, 297–311.

3. Charles Julien Brianchon, Solution de plusieurs problémes de géométrie, *Journal de l'École polytechnique*, Cahier 10, Tome 4, 1810, 1–15.

4. A. Cauchy, Du cercle tangent à trois cercles donnés, *Correspondance sur l'École polytechnique*, 1 (6) (July 1806): 193–195.

5. Lazare Carnot, *Géométrie de Position*, Paris: Duprat, An XI, 1803.

6. Lazare Carnot, *Essai sur la Théorie des Transversales*, Paris: Courcier, 1806.

7. Michel Chasles, *Traité de Géométrie Supérieure*, Paris: Bachelier, 1852.

8. Julian Lowell Coolidge, *A History of Geometrical Methods*, Oxford University Press, 1940.

9. Luigi Cremona, *Elementi di Geometria Projettiva*, Rome: G. B. Paravia, 1873.

10. Luigi Cremona, *Elements of Projective Geometry* 2nd edition, transl. by C. Leudesdorf, Oxford: Clarenden Press, 1893.

11. Leonhard Euler, *Introduction to Analysis of the Infinite, Book II*, 1748, (trans John Blanton), New York: Springer, 1990 (original 1748).

12. Leonhard Euler, De centro similitudinis [E693], *Leonhardi Euleri Opera Omnia* sub ausp. Soc. Scient, Nat. Helv. 1911- Series I vol. 26, 276–285 (original 1791).

13. L Gaultier, Mémoire sur les moyens généraux de construire graphiquement un circle déterminé par trois conditions et une sphère déterminée par quatre conditions, *Journal de l'École polytechnique* XVI, 1813, 124–214.

14. Gaspard Monge, *Géométrie descriptive. Leçons donnée aux Écoles normals, l'an 3 de la République*, Paris: Baudouin, an VII/1799.

15. J V Poncelet, Problêmes de géométrie, *Correspondance sur l'École polytechnique*, tome II, 1813, 271–274.

16. J V Poncelet, *Applications d'analyse et de géométrie qui ont servi, en 1822, de principal fondement au traité des propriétés projectives des figures, etc.*, 2 tomes, Paris: Mallet-Bachelier, 1862–64.

17. J V Poncelet, 7 cahiers de 1813–14, in [16], Tome 1, 1–441.

18. J V Poncelet, Essai sur les propriétés projectives des sections coniques, 1820, in [16], Tome 2, 365–441.

19. J. V. Poncelet, *Traité des Propriétés Projectives des Figures*, Paris: Bachelier 1822.

20. René Taton, *L'oeuvre scientifique de Monge*, Paris: Presses Universitaire de France, 1951.

21. John Wesley Young, *Projective Geometry*, Mathematical Association of America, Chicago: Open Court Publishing, 1930.

# Chapter 9
# Foci

## 9.1 Foci Before Poncelet

It is a formidable challenge to include foci in a projective treatment of the conic sections. They are, of course, not preserved under projection. Poncelet wrote in his 1822 *Traité*, Art. 446: "Although the properties of foci (*foyers*) ... seem to not be among those we have called *projectives*, ... they follow nevertheless in a very simple manner from foundational principles ...."

The earliest record we have of foci, although given no special name at the time, is in the *Conics* of Apollonius. The eccentricity property is not found in the *Conics* of Apollonius, but Pappus mentioned this property of conics in Book VII Prop 238, so it may have been know around the time of Apollonius, if not by Apollonius: Given a line $AB$ (directrix) and a point $C$ (focus) not on $AB$, and a given ratio, "Then point $D$ is on a conic," if the ratio $CD : DE$ equals the given ratio (eccentricity), " and if the ratio $= 1$ it is on a parabola, if $> 1$ on a hyperbola and if $< 1$ an ellipse." [8, p. 135]

Gregory of St. Vincent neatly summarized, in 1647, Apollonius's characterization of foci—called *poli* by St.Vincent: "the points at which the [major] axis is divided so the rectangle given by the [resulting] segments [$AF \cdot FB$ in Fig. 9.2] contains the quarter part of the *figure*, (i.e., equals the square on the semi-minor axis)." [7, p. 243] When we let $2a$ be the major axis and $2b$ the minor axis of an ellipse, this definition is equivalent to the modern $a^2 = b^2 + c^2$, when $c$ is the distance from the center to the focus. (The name *focus* is from Kepler, 1604, although a variety of names would be used for decades to come.)

La Hire had the benefit of minor advances by Kepler, Mydorge, and Gregory of St. Vincent. Mydorge, in the definitions at the beginning of his first book, gave the same definition as Apollonius for foci of the ellipse and hyperbola, but he also defined the focus of a parabola as the point inside the parabola, on the axis, at distance equal to a quarter of the *parameter*.

© Springer Nature Switzerland AG 2020
C. Baltus, *Collineations and Conic Sections*,
https://doi.org/10.1007/978-3-030-46287-1_9

**Fig. 9.1** Based on
Apollonius Book 3 Prop. 42
and La Hire Book 8 Prop.11,
1685

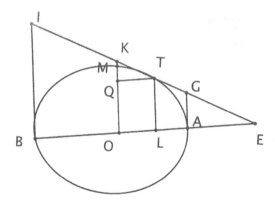

We begin with a Theorem, important to Apollonius and to La Hire in deriving the focus properties of Theorem 9.3. Our description is for the ellipse, but there is a corresponding lemma for the hyperbola.

**Theorem 9.1 (Apollonius Book 3 Prop. 42)** *Let $AB$ be the major axis of an ellipse with a tangent at $T$ meeting, at $G$ and $I$, the tangents at $A$ and $B$, respectively. (See Fig. 9.1.) Let $OM$ be the half-ordinate at center $O$. Then $OM^2 = BI \cdot AG$. So this product is independent of the choice of tangent to the ellipse.*

***Proof*** Let $TL$ be the ordinate on $T$ and $K$ the midpoint of $IG$. Since $H(BA, EL)$, then by a parallel projection $H(IG, ET)$. By Theorem 4.4, $IK^2 = KT \cdot KE = (KE - TE)KE$, so $TE \cdot KE = KE^2 - IK^2 = (KE + IK)(KE - IK) = IE \cdot GE$.

Then, treating $OK$ as the diameter with ordinate $QT$, we have $OM^2 = OQ \cdot OK$, so $OM^2 = LT \cdot OK$, where $TQ \parallel AB$ and $Q$ is on $OM$. Further, since $TE \cdot KE = IE \cdot GE$, then by parallels and similar triangles, $TL \cdot KO = IB \cdot GA$. Thus, $IB \cdot GA = TL \cdot KO = OM^2$. So the product $IB \cdot GA$ is independent of the choice of the tangent to the ellipse.                                                        $\square$

La Hire (Book 8 Prop. 7, 1685) and, later, Poncelet (Art. 447, 1822), characterized the foci, $F$ and $H$, for the ellipse and hyperbola by a construction process. We'll give the construction for the ellipse; there is a corresponding process for the hyperbola. The construction starts with any tangent to the ellipse. The following theorem is the claim that the construction yields the same points as foci no matter which tangent is selected.

**Theorem 9.2 (La Hire and Poncelet Foci)** *Let $AB$ be the major axis of an ellipse. We construct the circle with diameter $AB$, and then take any tangent, meeting the ellipse at $T$, meeting, at $G$ and $I$, respectively, perpendiculars to the axis at $A$ and $B$, and meeting the circle at $D$ and $C$. At $D$ and $C$ draw perpendiculars to the tangent, meeting the axis at $F$ and $H$. Then $H$ and $F$ are independent of the choice of the tangent. ($H$ and $F$ are the foci.)*

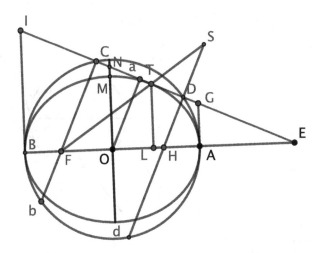

**Fig. 9.2** Based on La Hire's figure, Book 8 Prop. 7, 1685

***Proof*** See Fig. 9.2. By similar triangles,

$$\frac{BI}{CF} = \frac{BE}{EC} = \frac{ED}{EA} = \frac{DH}{AG}.$$

So $BI \cdot AG = CF \cdot DH$. By Theorem 9.1, $BI \cdot AG$ is a constant, $OM^2$, so $CF \cdot DH$ is a constant. Now, $CD$ is a chord of the circle, with perpendiculars $Dd$ and $Cb$ at its endpoints, so $Cd$ is a diameter. Thus, $CF = Hd$. This means $Hd \cdot DH$ is a constant; by the power of point $H$, $Hd \cdot DH = BH \cdot HA$ and this product equals the constant $OM^2$; only two points, $H$ and $F$, satisfy the equation.                                         □

Note that when we draw from center $O$ a perpendicular meeting the tangent at $a$, then $a$ is the midpoint of chord $CD$, so, by parallels, $O$ is the midpoint of $HF$.

Based on this discussion, we give the following definitions of the foci of conic sections, noting that Apollonius, La Hire, and Poncelet did not designate any statement as a "definition."

**Definition XVII**

i.  (Apollonius) Let an ellipse or hyperbola have major axis $AB$ and minor axis whose half-length is $b$. Then the two points $F$ and $H$ on the major axis satisfying $AF \cdot FB = b^2$ and $AH \cdot HB = b^2$ are the *foci*.
ii. (La Hire) Let an ellipse or hyperbola have major axis $AB$, or, in the case of parabola, axis with vertex $B$. Draw the circle with diameter $AB$ or, in the case of a parabola, the line $m$ on vertex $B$ perpendicular to the axis. Let any tangent to the conic section meet this circle at $C$ and $D$, or meet $m$ at $C$. Then the perpendiculars to the tangent at $C$ meet the major axis at a point $H$ which does not vary as $C$ varies on the conic, and $H$ is a *focus* of the conic. Likewise, for

**Fig. 9.3** Based on La Hire's figure, Book 8 Prop. 7, 1685

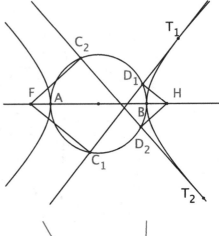

**Fig. 9.4** Based on La Hire Book Prop. 1 (parabola, focus $F$), 1685

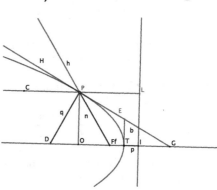

the ellipse and hyperbola, the perpendiculars to the tangent at $D$ meet the major axis at the second *focus*, $F$. See Fig. 9.3 for the case of the hyperbola.

What about the focus of the parabola? Although Apollonius did not consider a focus of a parabola, we saw it mentioned by Pappus as the conic for which the eccentricity is 1. Mydorge, St.Vincent, and La Hire all defined the focus to agree with a modern definition: the point on the axis of the parabola at distance $p$ from the vertex of a parabola defined by equation $y^2 = 4px$. $4p$ is the *parameter* as given by Apollonius.

La Hire's development agrees with that for the ellipse and hyperbola, as we'll see in the definition that follows. In place of the circle whose diameter is the axis of an ellipse or hyperbola, for the parabola we have the line, $m$, on the vertex of the axis that is perpendicular to the axis. He showed in Proposition 1 of Book VIII of 1685 [5] that the construction produces the same point, the focus, no matter which tangent we select (Fig. 9.4).

Three important properties of foci of the ellipse and hyperbola, known to Apollonius, follow.

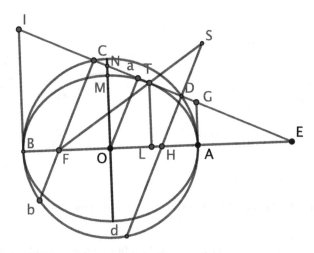

**Fig. 9.5** Based on La Hire's figure, Book 8 Prop. 7, 1685

### Theorem 9.3 (Apollonius Book 3 Prop. 45, 48, 52)

*a. Lines from the two foci to a point of tangency meet the tangent in equal angles.*
*b. For the ellipse, the sum of the lengths of these line segments is always the length*
*   of the axis.    (For the hyperbola, it is the difference of lengths.)*
*c. In the diagram of La Hire's Book 8 Prop. 7, 1685, I F (not drawn) is perpendic-*
*   ular to F G.*

***Proof in the Case of an Ellipse*** See, again, Fig. 9.2, repeated in Fig. 9.5, the figure
from La Hire Book 8 Prop. 7, 1685.

*a.* Let the perpendicular to $AB$ from $T$ meet $AB$ at $L$. We have harmonic sets:
   $H(EL, AB)$ and $H(ET, CD)$. By a parallel projection, $H(ET, GI)$. By similar
   triangles, and by the harmonic relationship,

$$\frac{CF}{DH} = \frac{CE}{DE} \quad \text{and} \quad \frac{CE}{DE} = \frac{TC}{TD}.$$

   So we have similar right triangles $TDH$ and $TCF$, so the lines from the foci to
   the point of tangency, $HT$ and $FT$, meet the tangent in equal angles.
*b.* We let $FT$ and $HD$, extended, meet at $S$. $\angle STD \cong \angle CTF \cong \angle DTH$. Since
   $TD \perp HS$, then $HT = ST$. But the diameter of the circle, $Db$, equals $FS$ since
   they are opposite sides of a parallelogram, and $FS = FT + TH$. So the sum
   $FT + TH$ is constant for any choice of tangent to the conic. (For the hyperbola,
   it is the difference of lengths which is constant.)
*c.* Since $BI \cdot GA = OM^2 = BF \cdot FA$, then $\triangle IBF \sim \triangle FAG$, so $IF \perp FG$.
   (Lines not drawn in the diagram)                                               □

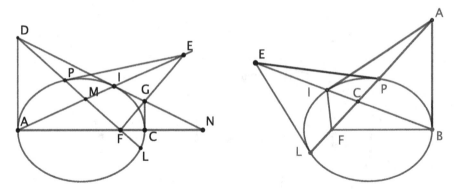

**Fig. 9.6** Based on La Hire's Book 8, 1685. Left Prop. 23; Right Prop. 24

Then we note two important theorems, La Hire's Prop. 23 and 24 of Book 8, 1685. Both seem to have been unknown before La Hire.

**Theorem 9.4 (La Hire Book 8 Prop. 23, 1685)** *Let E be a point on the polar of focus F with respect to a conic. Then EF is perpendicular to the polar of E (which lies on F.)*

***Proof*** See Fig. 9.6 Left. Given a conic with major axis $AC$, and from a point $E$ on the polar of $F$ with respect to the conic tangents are drawn, meeting the conic at $P$ and $L$. $F$ must be on $PL$. Let $PL$ extended meet $AD$, the tangent at $A$, in point $D$. Since $D$ is on the polar of $E$, then $E$ is on the polar of $D$, which meets the conic in points $A$ and $I$, and meets $PL$ in $M$. Let $DI$ meet $AC$ extended at $N$ and meet at $G$ the perpendicular at $C$ to $AC$. Now, $H(GD, IN)$ and $H(EM, IA)$. Since the harmonic sets have a common point, lines on the corresponding points are concurrent: $GE \cap DM \cap NA$ is $F$. So $G$ lies on $FE$. By Theorem 9.3c, $PL \perp FE$.
□

(There is a corresponding proof when the conic is a parabola.)
Finally,

**Theorem 9.5 (La Hire Book 8 Prop. 24, 1685)** *If tangents from an exterior point A meet a conic, with focus F, at I and B, then FA bisects angle IFB.*

The proof uses a lemma found in Desargues' work of 1639.

**Lemma (Harmonic Pencil with Perpendiculars)**
   *Suppose lines $a, b, c, d$ form a harmonic pencil, $H(ac, bd)$. Then c bisects the angle formed by b and d exactly when $a \perp c$.*

***Proof of Lemma*** Suppose that $c$ bisects the angle formed by lines $b$ and $d$. Take a line $m$ parallel to $a$, where lines $b, c, d$ meet $m$ in points $B, C, D$, respectively. $m$ and $a$ meet at infinity. Because the harmonic relation is preserved, $H(C\infty, BD)$, so $C$ is the midpoint of $BD$. Because, also, $c$ bisects the angle formed by $b$ and $d$, then $c \perp m$, so $c \perp a$. (Euclid Book 6 Prop. 3: The bisector of an angle of a triangle divides the opposite side in the ratio of the adjacent sides.) The converse follows in a similar way. □

***Proof of Theorem*** See Fig. 9.6 Right. Let line $FA$ meet the ellipse at $L$ and $P$. Let tangents at $L$ and $P$ meet at $E$. This means that $E$ is on the *polar* of $F$. By Prop. 23, $EF \perp FA$. Now, $H(EC, IB)$ where $EB$ meets $FA$ at $C$. So lines $FE$, $FI$, $FA$, and $FB$ are a harmonic pencil, or are *harmonicales*. But since $EF \perp FA$, then by the lemma $FA$ bisects $\angle IFB$. (A simple modern proof of Prop. 24 is in [1, p. 12].) □

## 9.2 Foci in Poncelet and Chasles

Poncelet arrived at a projective description of the focus of a conic section, which Chasles [2] and Cremona [3] would develop into a definition of the focus. A more advanced development can be found in Cremona's Chapter 23 of his [4] of 1893, a chapter not found in his 1873 original.

First we need a definition of a conic in terms of its focus. The following definition is the property noted by Pappus and is found in most calculus textbooks. Its equivalence to other definitions is usually shown by derivation of the $x$, $y$ equation of the conic.

**Definitions XVIII** Given a point, $F$, the *focus*, and a line, the *directrix*, not on $F$, and a positive constant $k$, then the set of points, $P$, such that the distance $PF$ is $k$ times the distance from $P$ to the directrix, is a conic section. The conic is an ellipse, parabola, or hyperbola as $k$ is less than 1, equal to 1, or greater than 1, respectively. And all conics are found this way.

We have already stated that a conic is the image of a circle under a central collineation. We will show that, at least for a certain choice of central collineation, the image of a circle satisfies the definition above of a conic section. Further, Poncelet's work reaches a new description of a focus, a description which Chasles and Cremona would adapt as the definition of a focus. We will reach Poncelet's conclusion by a new method.

Let $F$ be a point and the center of the circle, $c'$, that lies on a given point $T'$, and let $M'$ and $N'$ be points where neither $F$ nor $T'$ is on line $M'N'$. Let $T$ be a point on line $FT'$. Let $\phi^{-1}$ be the central collineation with center $F$, axis $M'N'$, and which maps $T'$ to $T$. See Fig. 9.7, from Poncelet. (We have chosen to call the collineation $\phi^{-1}$, since it is the inverse mapping that will appear in Poncelet's theorem.)

What do we note about the image, $c$, of the circle $c'$?

1. The tangent to $c'$ at $T'$ meets its image, the tangent to $c$ at $T$, at point $P'$, on axis $M'N'$.
2. Let $q$ denote the line on $F$ that is perpendicular to $M'N'$. Then curve $c$ is symmetric about line $q$. Why? Let $V'$ (not in Poncelet's diagram) be the reflection of $T'$ over line $q$, and let $V$ denote the image of $V'$ by collineation $\phi$. Where is $V$?

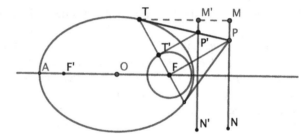

**Fig. 9.7** Based on Poncelet's
*Fig. 69*, 1822: Focus and its
polar

It lies on line $FV'$. Now, the vertical line on $T'$ meets the axis of the collineation
at infinity, so that vertical line is mapped to the parallel line on $T$, so $V$ is the
reflection of $T$ over line $q$.

3. Let $P$ be the point on line $TP'$ that lies on the perpendicular, at $F$, to line $TF$.
   Then $P$ is the image under $\phi^{-1}$ of the point at infinity of line $T'P'$. Why? $P'$ is
   fixed and $T'$ is mapped to $T$, so the image of that point at infinity lies on $TP'$.
   Further, $F$ is the center of collineation $\phi^{-1}$, so the image of that point at infinity
   lies on the line on $F$ that is parallel to $T'P'$. So the image is $P$.

Now we consider the inverse collineation to $\phi^{-1}$, $\phi$, mapping conic $c$ to circle $c'$.
Since $P$ is the image under $\phi^{-1}$ of a point at infinity, then $P$ lies on the vanishing
line of $\phi$. So line $MN$ in the diagram, which lies on $P$ and is parallel to $M'N'$, is the
vanishing line of $\phi$. (We assume that $M$ and $M'$ are placed so line $MM'$ is parallel
to $q$.)

Because of parallels in Poncelet's diagram,

$$\frac{FT}{FT'} = \frac{TP}{PP'} = \frac{M'T}{MM'}, \quad \text{so} \quad \frac{FT}{M'T} = \frac{FT'}{MM'}.$$

Now, $FT'/MM'$ is a constant, independent of the choice of $T'$ on the circle.
This means that the points $T$ are on the conic with focus $F$, directrix $MN$, and
$k = FT'/MM'$.

We observe that line $MN$ is the polar of point $F$ with respect to the conic, $c$.
Why? With respect to circle $c'$, the center, $F$, is the pole of the line at infinity. The
pole/polar relation is preserved under a central collineation, $\phi^{-1}$—which maps the
line at infinity to $MN$ and $F$ to itself.

Poncelet started with La Hire's definition of foci, as in Theorem 9.2. He made
important use of La Hire's Book 8 Prop. 23, although he had his own proof (Arts.
433 and 460). He devoted articles in [6] to pairs of conics that share a common
secant, and/or a focus. With his *Fig. 69*, in our *Fig. 9.7*, he examined a circle whose
center is a focus, $F$, of the conic. He showed the following:

**Theorem 9.6 (Poncelet Focus Theorem, Art. 457, 1822)** *Let $c$ be a conic with
focus $F$. Let $MN$ be the polar of $F$. We take the central collineation with center $F$
and vanishing line $MN$, whose axis is to be designated later. Then the image of $c$ is
a circle with center $F$.*

**Proof** The discussion above proves this theorem, except for the requirement that for every $P$ on vanishing line $MN$ we need $PF$ to be perpendicular to the polar of $P$—which must lie on $F$. But this is exactly the property of conics detailed in La Hire's Prop 23 of Book 8 of [5]: $TF \perp PF$ and these are conjugate lines at $F$ of conic $c$. Under $\phi$ they become conjugate diameters of $c'$, as each is mapped to itself. Since $TF \perp PF$ and $F$ is the center of $\phi$, then $TF$ and $PF$, and all pairs of conjugate diameters of $c'$, are perpendicular. So $c'$ is a circle with center $F$. □

Our discussion, in fact, proves more, which we state in a theorem, which includes La Hire's Prop. 23 of Book 8 of 1685.

**Theorem 9.7** *Suppose that $F$ is a point on the axis of a conic section so that any $P$ on the polar of $F$ is joined to $F$ by a line perpendicular to the polar of $P$. Then $F$ is a focus of the conic section, with the conic section satisfying the focus-directrix definition of a conic section, as in Definition XVIII.*

*Conversely, in a conic section with focus $F$, any point $P$ on the polar of $F$ is joined to $F$ by a line perpendicular to the polar of $P$.*

It was Poncelet's successor, Michel Chasles, who took the next step in redefining the focus of a conic. But before we turn to Chasles, we will note an additional applications by Poncelet, in Art. 462, of his discovery concerning a circle whose center is a focus of another conic.

**Theorem 9.8 (Poncelet's Art. 462, 1822)** *Suppose (Poncelet's Fig. 71, our Fig. 9.8) we have a circle with center $F$, an outside point $C$, and tangents from $C$ meeting the circle at $T$ and $T'$. Then a tangent to the circle is drawn meeting $CT$ at $A$ and meeting $CT'$ at $B$. The first claim is that $m\angle AFB$ is independent of the choice of tangent $AB$. Second, suppose we have a conic with focus $F$ and outside point $C$, and tangents from $C$ meeting the conic at $T$ and $T'$, and a tangent to the conic meeting $CT$ at $A$ and meeting $CT'$ at $B$. Then $m\angle AFB$ is, again, independent of the choice of the tangent to the conic.*

**Proof** As in the diagram, we draw $FA$ and $FB$. Let $K$ (not marked in the diagram) be the point at which tangent $AB$ meets the circle. By a tangent property, $\angle TFA \cong \angle KFA$ and $\angle T'FB \cong \angle KFB$. So $m\angle AFB = m\angle T'FT/2$, a constant value. Now,

**Fig. 9.8** Poncelet's Art. 462: Invariance of $\angle AFB$

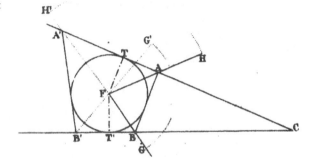

second, consider a conic with focus $F$ and outside point $C$, and line $MN$ the polar of $F$. Suppose there are tangents from $C$ meeting the conic at $T$ and $T'$, and a tangent to the conic meeting $CT$ at $A$ and meeting $CT'$ at $B$. When we apply a central collineation with center $F$ and vanishing line $MN$, the image is a circle with center $F$. Because the center of the collineation is $F$, $m\angle AFB$ is the same in the image. By the first part of the theorem, $m\angle AFB$ is, again, independent of the choice of the tangent to the conic.                                                                 □

Michel Chasles (1793–1880) developed La Hire's Book 8 Prop. 23, 1685, and the central collineation of Poncelet's Focus Theorem, Theorem 9.6, into a new definition of the foci.

**Definitions XIX (Chasles' Definition of Focus, *Traité des Sections Coniques*, 1865, Art. 275–276.)**
A focus of a conic is defined to be a point $S$ when a central collineation with center $S$ and with the polar of $S$ as the vanishing line maps the conic to a circle. See also [4, Cremona, Art. 341].

Thus, we have reached a definition of the focus based on a central collineation. After the next chapter, we will briefly look at Desargues' work with foci.

# References

1. A. V. Akopyan and A. A. Zaslavsky, *Geometry of Conics*, transl. A. Martsinkovsky, Providence, RI: American Mathematical Society, 2007.
2. Michel Chasles, *Traité des Sections Coniques*, Paris: Gauthiers-Villars, 1865.
3. Luigi Cremona, *Elementi di Geometria Projettiva*, Rome: G. B. Paravia, 1873.
4. Luigi Cremona, *Elements of Projective Geometry* 2nd edition, transl. by C. Leudesdorf, Oxford: Clarenden Press, 1893.
5. Philippe de La Hire, *Sectiones Conicae en novem libros distributae*, Paris 1685; French translation by Jean Peyroux, *Grand Livre des Sections Coniques*, Paris: Blanchard, 1995.
6. J. V. Poncelet, *Traité des Propriétés Projectives des Figures*, Paris: Bachelier, 1822.
7. P. Gregorii a Sto. Vicentio, *Opus geometricum quadraturae circuli et sectionum coni*, Antwerp: Ioannem and Iacobum Meursios, 1647.
8. J. H. Weaver, Pappus. Introductory paper, *Bulletin of the American Mathematical Society*, vol 23, No. 3, 1916, 134–135.

# Chapter 10
# Steiner: Cross-Ratio, Projective Forms, and Conics

The most important works in projective geometry following closely on Poncelet's 1822 work were A. Moebius's *Der Barycentrische Calcul* (1827) and Jacob Steiner's *Systematische Entwickelungder Abhängikeit geometrischer Gestalten von einander* (1832). Moebius introduced the term *collineation* and studied collineations of the plane by means of barycentric coordinates, a type of homogeneous coordinates that we will examine in Chap. 13.

Jacob Steiner (1796–1863) moved from his family farm in Switzerland to attend school only at age 18. But that school was the Pestalozzi School, and Steiner took to heart the emphasis on independent thinking. After that limited education, he spent more than a decade supporting himself by tutoring and some teaching, attending university lectures and writing papers. His search for deeper bases for mathematics, and for unity and coherence, led to his book of 1832.

## 10.1  Cross-Ratio and Projective Forms

The *cross-ratio* among four collinear points appeared, with no special name, in [1, Brianchon 1817] and, with the name *Doppelschnittverhaltniss*, in [10, Moebius 1827]. It became especially important with Steiner. The dual is the *cross-ratio* of four concurrent lines.

**Definitions XX**  If $A, B, C, D$ are collinear, the *cross-ratio*, which we designate

$$CR(AC, BD), \text{ is } \frac{AB \cdot CD}{AD \cdot CB}$$

where the lengths are signed.

For four lines $a, b, c, d$, concurrent at a point $P$, the *cross-ratio* $CR(ac, bd)$ is $CR(AC, BD)$ when some line not on $P$ meets the lines, respectively, at $A, B, C, D$.

© Springer Nature Switzerland AG 2020

C. Baltus, *Collineations and Conic Sections*,

https://doi.org/10.1007/978-3-030-46287-1_10

This definition makes sense since, by Theorem 10.2, the cross-ratio of four collinear points is invariant under projection.

We note that the harmonic relation among four collinear points holds when the cross-ratio is $-1$.

By solving an equation, we see

**Theorem 10.1** *Given three collinear points $A$, $B$, $C$ and a value, $k$, other than $0$, $1$ or $\infty$, then there is a unique point $D$ collinear with $A$, $B$, $C$ so $CR(AC, BD) = k$.*

Important properties of the cross-ratio are contained in this theorem. The first two are immediate consequences of the definition of the cross-ratio.

**Theorem 10.2 (Cross-Ratio Theorem)** *For collinear points $A$, $B$, $C$, $D$,*

*a. $CR(AC, BD) = CR(CA, DB)$,*
*b. $CR(AC, BD) = CR(BD, AC)$*
*c. the cross-ratio is invariant under projection.*

The third property can be shown by Carnot's Principle, as expressed in Poncelet's Art. 20 of 1822, our Theorem 8.3. Carnot's Principle applies since the cross-ratio depends only on sines of the angles formed when the four points are joined to the center of a perspectivity. This was the proof presented by Steiner [13, Article 4]. See also [1, p. 7], and [12, Articles 21–27]. Pappus proved the invariability of the cross-ratio (Book 7 Prop. 129) under projection, without giving any particular name or context to the relation. But it was Steiner who made the cross-ratio, and the fact that it is preserved under projection, central to his projective geometry.

Steiner rebuilt the subject of projective geometry not in changing the problems it is to solve, but in the foundations on which the theory is built. Poncelet had turned projection from a tool in the study of conic sections into the main object of study. He wished to discover the properties of figures which are invariant under a projection, and used the results of that examination to solve problems both old and new. Now Steiner further broke down the process of projection. He began with *projective forms* and *projectively related forms*. In the plane, a projective form is either the set of points of a line, including its point at infinity, or a pencil, the collection of lines on a particular point.

**Definitions XXI** The *projective forms* in a plane are the points of a line, now called a *range*, and the *pencil—Strahlbüschel* for Steiner, a collection of lines on a point. That point is the *center*, or *Mittelpunkt*.

The elementary relations between projective forms are called *perspectivities*. The line-to-line perspectivity, say from line $l_1$ to line $l_2$ with center $P$ is the mapping that pairs each $X_1$ on $l_1$ to $X_2$ on line $l_2$ when $P$, $X_1$ and $X_2$ are collinear. See Fig. 10.1 Left.

The pencil-to-pencil perspectivity between pencils on $X$ and $Y$ is achieved by means of a line, $p$, on neither $X$ nor $Y$, which Steiner called a *Durchschnitt*: line $a$ on $X$ and $a_1$ on $Y$ are paired exactly when those lines meet on the designated line $p$. See Fig. 10.1 Right.

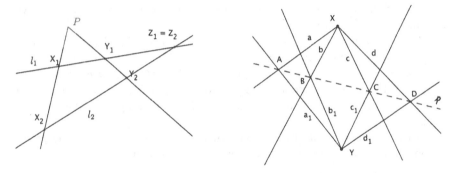

**Fig. 10.1** Left: Line-to-line perspectivity. Right: Pencil-to-pencil perspectivity

Further, there is a pencil-and-line perspectivity. Given a pencil with center $P$ and a line $m$ not on $P$, then every point $M$ of $m$ corresponds to line $MP$ of the pencil (Steiner p. 241).

Two pencils, or two lines, or a pencil and a line related by a composition of perspectivities are *projectively related*.

With these definitions, we have a corollary to Theorem 10.2c:

**Corollary** *For any two projectively related forms, the cross-ratio of any four elements—lines or points–of one form equals the cross ratio of the corresponding elements of the second form.*

We see in Fig. 10.1 that line-to-line and pencil-to-pencil perspectivities are effected by a central collineation that is applied to a line or a pencil.

Then comes a question that arises with the concept of projectively related forms: given a one-to-one pairing of the elements of two forms, when can we say that they are projectively related? We know that the cross-ratio of any four elements of a projective form agrees with the cross-ratio of the corresponding elements of a form in perspective. This leads to Steiner's answer, which can be regarded as a theorem or an alternative definition:

**Definition XXII** Two lines, or two pencils, or a line and pencil are *projectively related* under a correspondence of points or lines exactly when the cross-ratio of any four points or lines of one equals the cross-ratio of the corresponding four points or lines. A line or pencil may be projectively related to itself.

In the words of Steiner [13],

If the elements a, b, c, d, …and $a, b, c, d, \ldots$ of two figures [*Gebilde*] $\mathcal{A}$ and $\mathcal{B}$ are so paired in sequence that any four elementpairs [give equal corresponding cross-ratios], then the figures are projective [*projectivisch*] in relation to all the pairs of elements. [p 248, first Section, Article 6 $(\gamma)$]

Since three elements of a projective form, $A, B, C$, and a value $k$ other than 0, 1 or infinity, determines a fourth element $D$ so $CR(AB, CD) = k$, then pairing three

elements of one form with three elements of a second form determines a projective
relation between the forms.

We can map any $A, B, C$ of one form to any $A_1, B_1, C_1$ of a second form,
respectively, by a composition of two perspectivities. (See Exercise 6.) In the case of
two lines, we could first project, say, line $ABC$ to line $AC_1$, where $B^*$ denotes the
image, on $AC_1$, of $B$, then project line $AC_1$ to line $A_1C_1$ from center of projection
$B^*B_1 \cap CC_1$. (If line $ABC$ equals line $A_1B_1C_1$, then a third perspectivity may be
required.)

La Hire's Lemma 7 of 1673 can be extended, by essentially the same proof,
beyond the case of harmonic sets which have a point in common:

**Theorem 10.3** *When the intersection point of two projectively related lines is
related to itself, then the lines are in perspective. And the dual, likewise, holds,
namely, when two projectively related pencils share a line related to itself, then the
corresponding lines meet in collinear points.*

With the duality in the projective forms, duality in propositions was emphasized
by Steiner, as in listing the corresponding dual theorems in two columns. Gergonne
had done this in an article of 1827–1828 [8]. Steiner's use of the cross-ratio as the
defining characteristic of a projective relation would be followed by Michel Chasles
in his influential *Traité de Géométrie Supérieure*, of 1852. Chasles called the cross-
ratio the *rapport anharmonique*, with the harmonic relation and involution defined
in terms of the cross-ratio.

These innovations effected a change in the subject, moving the emphasis from
the study of plane-to-plane transformations to mappings from a line or pencil to a
line or pencil.

The novel nature of Steiner's work is evident in the Cross-Joins Theorem, [13,
Art. 24*b*]. Both the concept and the proof seem unlikely before Steiner.

**Theorem 10.4 (Cross-Joins Theorem)** *Suppose there are two projectively related
lines, $a$ and $a_1$, meeting in a point $D$ which is not necessarily paired with itself in
the projective relation, where points $A, B, C, \ldots$ lie on $a$, while the corresponding
$A_1, B_1, C_1, \ldots$ lie on $a_1$. Then the cross-joins, namely, $AB_1 \cap A_1B, AC_1 \cap
A_1C, BC_1 \cap B_1C, \ldots$, are collinear.*

***Proof*** We follow the strategy of Steiner. See Fig. 10.2 Left. We find an additional
pair of corresponding points, pairing a particular point $D$ on line $a$ with $D_1$ on $a_1$
so that $\{A, B, C, D\}$ is projectively related to $\{A_1, C_1, B_1, D_1\}$. We can let $D$
be the intersection of lines $a$ and $a_1$. We choose $D_1$ on line $a_1$ corresponding in the
projective relation to $D$ of line $a$: take $D_1$ so

$$CR(A\,C, B\,D) = CR(A_1\,C_1, B_1\,D_1).$$

This means the pencil $\{AA_1,\ AB_1,\ AC_1,\ AD_1\}$ on $A$ is projectively related to
the pencil $\{A_1A,\ A_1B,\ A_1C,\ A_1D\}$ on $A_1$. Further, line $A_1A$ in that projective
relation is paired with itself. So by Theorem 10.3, the pencils are in perspective.

The corresponding pairs meet in $G_1 = AB_1 \cap A_1B$, $G_2 = AC_1 \cap A_1C$, and $D_1 = AD_1 \cap A_1D$. Thus, $G_1, G_2$, and $D_1$ are collinear. We find, in the same way, projectively related pencils on $B$ and $B_1$, namely, $\{BA_1, \ BB_1, \ BC_1, \ BD_1\}$ on $B$ and $\{B_1A, \ B_1B, \ B_1C, \ B_1D\}$ on $B_1$. As before, we have collinear points $G_1 = AB_1 \cap A_1B$, $G_3 = BC_1 \cap B_1C$, and $D_1 = BD_1 \cap B_1D$. The two sets of three collinear points found share points $G_1$ and $D_1$, so all four points are collinear. We can continue finding projectively related pencils on $C$ and $C_1$, etc., showing that all the cross-joins are collinear.                                                    □

**Example 1** Here is a construction whose justification illustrates the power of Steiner's projective pairing of a line and a pencil. It is attributed to Chasles, taken from L. Wayland Dowling's [7, p 53]. See Fig. 10.2 Right. Let lines $u$ and $u_1$ meet at $O$, and let $S$ and $S_1$ be two points collinear with $O$. A line rotates about a point $U$, meeting $u$ and $u_1$ in points $A$ and $A_1$, respectively, as $A$ varies on $u$. Show that the locus of the intersections of $SA$ and $S_1A_1$ is a straight line. The strategy is to show that the pencils on $S$ and $S_1$ are projectively related, with the line $SS_1$ corresponding to itself. Thus the correspondence would be a perspectivity, whose corresponding lines meet on a line.

Solution: We apply Theorem 10.3. The points of lines $u$ and $u_1$ are in perspective from $U$, so projectively related, with $A$ paired with $A_1$, etc. Since the points of $u$ and $u_1$ are in perspective, respectively, with pencils on $S$ and $S_1$, then those pencils are projectively related. Line $SS_1$ is related to itself, so those pencils are in perspective. It follows that the corresponding lines meet in collinear points.                         □

An application of the Cross-Joins Theorem is Pappus's Hexagon Theorem, proved earlier, by a projection, at the end of Chap. 4, in Exercise 2.

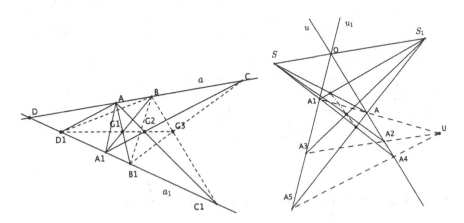

**Fig. 10.2** Left: Proof of cross-joins theorem based on Steiner's 1832 *Fig.* 33. Right: Solution of Example 1

**Fig. 10.3** Pappus's Hexagon
Theorem

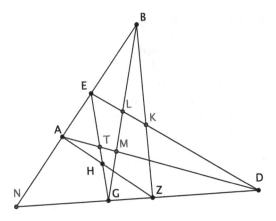

**Theorem 10.5 (Pappus's Hexagon Theorem)** *Let points $A, B, C$ lie on a line $a$ and points $A_1, B_1, C_1$ lie on line $a_1$. Then points of intersection of opposite sides of hexagon $AB_1CA_1BC_1$ are collinear.*

***Proof*** Take the projectivity that maps $A, B, C$ to $A_1, B_1, C_1$, respectively. By the Cross-Joins Theorem, points $AB_1 \cap A_1B$, $AC_1 \cap A_1C$, $BC_1 \cap B_1C$ are collinear. These three points are the points of intersection of opposite sides of hexagon $AB_1CA_1BC_1$. □

Pappus had proved this theorem in Book 7 of his *Mathematical Collection*. Using triangle similarity, Pappus showed that the cross-ratio of four collinear points is preserved under a line-to-line projection. Restating Pappus's proposition in terms of cross-ratios, he proved Theorem 10.3, that when two lines meet at $A$, and for points $B, C, D$ on one line and points $B_1, C_1, D_1$ on the second line we have $CR(AC, BD) = CR(AC_1, B_1D_1)$, then there is a projection from an outside point sending $B, C, D$ to $B_1, C_1, D_1$, respectively.

Here, in terms of cross-ratios, is Pappus's proof of his Hexagon Theorem, [11, p 272], using Fig. 10.3, adapted from [11, p 699, Pappus's Prop. 139].

Consider hexagon $DAZBGE$. By projection from $A$, $CR(GT, EH) = CR(GD, NZ)$. By projection from $B$, $CR(LD, EK) = CR(GD, NZ)$. So, $CR(GT, EH) = CR(LD, EK)$. This means corresponding points are joined by concurrent lines. Those concurrent lines meet at $M$, so $H, M, K$ are collinear. □

## 10.2   Conics with Steiner, Chasles, and Cremona

Steiner saw that a conic was the set of points of intersection of two projectively related pencils (see [6, p 202]). Suppose $a, b, c, B, B_1$ are points on a circle. Then the inscribed angles that the chords from corresponding chords $a, b, c$ form at $B$ are equal to the angles that the corresponding chords $a_1, b_1, c_1$, from $a, b, c$, form at

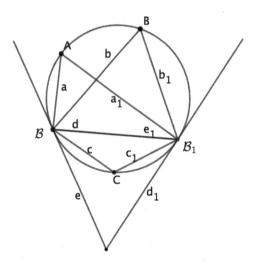

**Fig. 10.4** Steiner's *Fig.* 37, 1832, definition of a conic

$\mathcal{B}_1$. See Steiner's [13, Art. 37, Fig. 37], in our Fig. 10.4. Under this correspondence, since cross-ratios are equal, the pencils on $\mathcal{B}$ and $\mathcal{B}_1$ are projectively related and that relation is preserved under a projection that sends the circle to a conic. This leads to Steiner's new characteristic of a conic section. We state it as a definition, and we include the dual, a *line conic*.

**Definition XXIII** A *point conic* is the set of points of intersection of two projectively related pencils. A *line conic*, defined by its set of tangent lines, is the set of lines on corresponding points of two projectively related lines.

How did Steiner introduce the line conic? It is in Articles 37 and 38 of [13, 1832]. Looking at our Fig. 10.5, from Poncelet, 1822, the tangents $AB$ and $A'B'$, for different positions of $A$ and $A'$, are the lines joining the projectively paired points of lines $AC$ and $BC$. Steiner gave a proof, involving the cross-ratio, of the projective relation of the two lines. But we go back to Theorem 9.8, by Poncelet, with Poncelet's *Fig.* 71 of 1822, in our Fig. 10.5.

Let $F$ be the focus of a conic, with $CT$ and $CT'$ tangent to the conic. We say a point $A$ on $CT$ corresponds to point $B$ on $CT'$ when line $AB$ is tangent to the conic. That correspondence is projective. By Theorem 9.8, of Poncelet, $m\angle AFB$ is independent of the choice of the tangent to the conic. Line $FA$ of the pencil on $F$ is paired with line $FB$ of the same pencil when $AB$ is tangent to the conic, and that pairing is a rotation by the constant angle $\angle AFB$. Now, a rotation is the composition of two reflections over lines on the center of the rotation, and reflections are central collineations, so the pencils described are projectively related. This means point $A$

**Fig. 10.5** Poncelet's Art.
462: invariance of $\angle AFB$

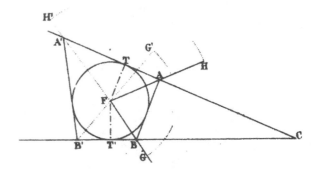

on line $AC$ is paired projectively with point $B$ on line $T'B$. The set of lines $AB$ form the line conic.

J.L. Coolidge, a century later, agreed that a circle (and a conic) is the set of intersections of corresponding lines in projectively related pencils, but "I find no proof that every curve so generated is the projection of a circle" [3, p 97]. In other words, given a locus of points of intersection of corresponding lines of two projectively related pencils, how do we know it is the image of a circle under a central collineation? We turn to Luigi Cremona's [4, Article 114], with his *Fig. 82*, and note a corresponding argument by Chasles [2, Art. 8] from 1865. Cremona explicitly referred to a central collineation.

**Theorem 10.6 (Chasles, Cremona)** *Suppose that pencils on points $Q$ and $A$ are projectively related, with corresponding lines meeting on a locus including $Q, A, B, P$. (Also, assume that line $AQ$ does not correspond to itself, since otherwise, by the dual of La Hire's Lemma 7, then the pencils are in perspective and the corresponding lines meet in a line.) Then the set of points of intersection of corresponding lines is the image of a circle under a central colliineation.*

***Proof*** See Fig. 10.6. Let line $AQ$ of pencil $A$ correspond to line $q$ of pencil $Q$. Draw a circle tangent to $q$ at $Q$, and let this circle meet $AQ$ at point $A'$. We first relate the pencil on $A'$ to the pencil on $Q$ as follows: To a line on $Q$ meeting the given locus at $P$ we correspond line $A'P'$ where line $QP$ meets the circle at $P'$. So a line $QB$ is paired with line $A'B'$ defined in the corresponding way. Since $Q, A', P', B'$ are all on a circle, $\angle P'QB' = \angle P'A'B'$, and so, with each line $QD'$ corresponding to the line $A'D'$ that it meets on the circle, pencils $Q$ and $A'$ are projectively related. So pencils $A'$ and $A$ are

projectively related. Further, line $AA'$ of pencil $A$ corresponds to itself, line $A'A$ of pencil $A'$. Now, by Theorem 10.3, if two projectively related pencils have the line on the two vertices corresponding to itself, then the pencils are in perspective. Therefore corresponding lines of pencils $A$ and $A'$ meet on a line $s$. So, finally, consider the central collineation with center $Q$, axis $s$, and which maps $A'$ to $A$. Each point, such as $B'$, on the circle is mapped by the straightedge construction of

**Fig. 10.6** Cremona's *Fig.* 82, 1893

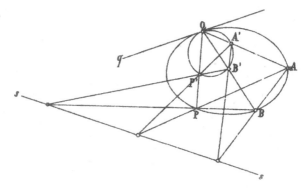

a central collineation to the corresponding point $B$ on the locus. Thus, the locus is the image of a circle under a central collineation, and so the locus is a conic. ☐

## 10.3 Constructions of Conics

Steiner's characterization of a point conic as the set of intersections of corresponding lines of two projectively related pencils, or of a line conic as the dual, leads to some interesting and pretty constructions.

The first is from Colin Maclaurin, in a 1735 article [9, p 152], although originally presented nearly 20 years earlier, far earlier than Steiner's work, which we use to explain the construction. See Fig. 10.7 Lower Center. We are given two lines, $AQ$ and $AR$, and three non-collinear points, $D, C, S$, none on the two lines. We draw a line on $D$, meeting the two given lines in $Q$ and $R$, respectively. This establishes a projective correspondence of the two given lines, pairing points $Q$ with points $R$. Then we draw the lines $CR$ and $SQ$ on the corresponding points $R$ and $Q$; this establishes projectively related pencils, one on $C$ and one on $S$. The points of intersection of the corresponding lines, including point $P$ in the diagram, form a conic. Note that if the initial line on $D$ should be $DC$, then the point $F = AQ \cap CD$ will lie on the conic. In a similar way, points $E$ and $C$ and $S$ and $A$ in the diagram are on the conic. Thus we have a method of finding as many points as desired on the conic lying on five given points, no three collinear.

This corollary about a unique conic on five given points $A, F, C, S, E$ follows Steiner's definition. The pencil $AC, AS, AE$ on $A$ is projectively related in exactly one way with the pencil $FC, FS, FE$ on $F$, with Maclaurin's construction providing a method of finding more points on the conic [5]. When three of the five points are collinear, then the resulting conic section consists of two lines, or even one line.

The construction serves as proof for an important theorem:

**Theorem 10.7 (Five Points)** *For any given five points, there is a unique conic section on the five points.*

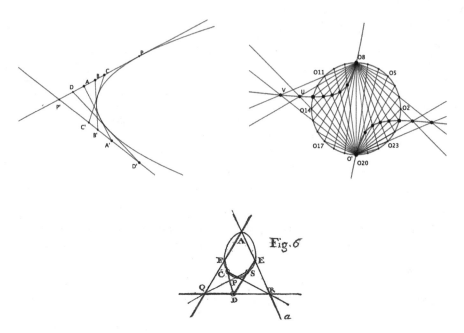

**Fig. 10.7**  Left: Cremona *Fig.* 100, 1893. Right: *What is Mathematics?* 1941, p. 205. Lower center: Maclaurin's Construction of a Conic, 1735

The second example, in Fig. 10.7 Left, from Cremona, 1893, is a line conic. If one marks equally spaced points on two intersecting lines and labels the points in order, in either direction, starting anywhere, the points of the lines will be projectively related, since translation and dilation are projective mappings.

Figure 10.7 Right is from *What is Mathematics?* The 24 points equally spaced around the circle give projectively related pencils (Why?) with centers $O$ and $O'$, producing a point conic.

These are just three examples, selected for their simplicity, from hundreds of methods and mechanical devices for point-by-point tracing of conics sections.

## 10.4   Excursion: Central Collineations and Perspectivities

What can we say about the connection of central collineations and projectivities?

Clearly, any line-to-line perspectivity of a line $m$ to a line $m'$, can be achieved by a central collineation whose center is the center of the perspectivity and which maps $m$ to $m'$. Likewise, a pencil-to-pencil perspectivity, where each line of a pencil $M$ is paired with the line of pencil $M'$ that it meets on a designated line $p$, is achieved by a central collineation with axis $p$ which maps point $M$ to point $M'$.

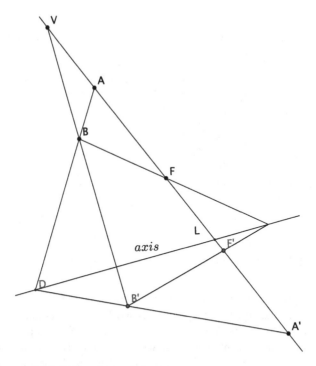

**Fig. 10.8** Theorem 10.9. Collineation as projectivity

Conversely, any mapping by a central collineation of a line $m$ to a line $m'$ (or of a pencil $M$ to a pencil $M'$) is a line-to-line perspectivity (or pencil-to-pencil perspectivity). With the following theorem, we see that the mapping of a line to itself by a central collineation is a projectivity. The dual theorem will, of course, hold.

**Theorem 10.8** *Let $\phi$ be a central collineation with center $V$, axis $l$, not on $V$, that maps $A$ to $A'$. Then the mapping produced of line $VA$ to itself is the composition of two perspectivities.*

***Proof*** See Fig. 10.8. Let the central collineation have center $V$, axis $l = LD$, and map $A$ to $A'$. Here is the construction that maps another point $F$ of $VL$ to its image: take a line $VB$, $B$ not on the axis, on $V$. To find $B'$, we draw line $AB$ meeting the axis at $D$. Then, projecting line $VA$ to $VB$ from center $D$, $B'$ is $DA' \cap VB$. Now, $F'$ is found by projecting line $VB$ back to $VA$, but where the center is the point where the axis, $l$, meets line $BF$. $F'$ is the image of $B'$ under this projection. Thus the central collineation on line $VA$ is the composition of these two perspectivities.

□

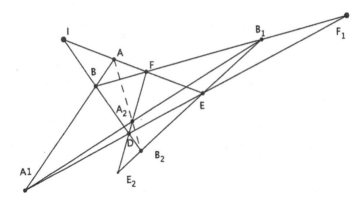

**Fig. 10.9** Proof of the triangle-on-a-triangle theorem based on Steiner's 1832 *Fig.* 30.

## 10.5   Notes and Exercises

**Note 2**  In the projective correspondence of lines on $B$ and $B_1$, in our Fig. 10.4, Steiner's *Fig.* 37, what about the tangents at $B$ and $B_1$? Consider a point, $e$, moving along the conic, approaching $B$. For lines in the pencil on $B$, as $e$ approaches $B$, the lines $Be$ approach the tangent at $B$. Meanwhile, the lines $B_1e$ in the pencil on $B_1$ approach line $BB_1$. So in the one-to-one correspondence of the pencil on $B$ with the pencil on $B_1$, the tangent at $B$ is paired with line $BB_1$.

**Note 3**  Steiner's Article 23, of 1832, is an interesting example in which composition of transformations is applied.

**Theorem 10.9 (Steiner's   Triangle-on-a-Triangle   Theorem)** *Let* $\triangle DEF$ *be inscribed in the triangle with vertices* $B$, $B_1$, *and* $B_2$, *as pictured in Fig. 10.4. Let* $BB_1 \cap EF = I$, $B_1B_2 \cap DF = E_2$, *and* $BB_2 \cap DE = F_1$. *We consider the composition of three line-to-line perspectivities:* $\phi$ *with center* $B$ *mapping line* $EF$ *to line* $EF_1$, *followed by* $\phi_1$ *with center* $B_1$ *mapping line* $EF_1$ *to line* $E_2F$, *followed by* $\phi_2$ *with center* $B_2$ *mapping line* $E_2F$ *back to line* $EF$. *Then*

(a)  *the composition* $\phi_2 \circ \phi_1 \circ \phi$ *is the identity mapping line* $EF$ *to itself, and*
(b)  *if* $A$ *is any point of line* $EF$, *then* $A$ *will be mapped to a point* $A_1$ *on line* $DE$, *which will be mapped to a point* $A_2$, *on* $DF$, *and that* $A_2$ *will be mapped to the original point* $A$, *meaning that* $A$, $A_2$, *and* $B_2$ *are collinear.*

***Proof*** $\phi$ maps points $E$, $F$, $I$ to, respectively, $E$, $F_1$, $D$. Then $\phi_1$ maps $E$, $F_1$, $D$ to, respectively, $E_2$, $F$, $D$. And then $\phi_2$ maps $E_2$, $F$, $D$ to, respectively, $E$, $F$, $I$. Therefore the composition $\phi_2 \circ \phi_1 \circ \phi$ is a projectivity of line $EF$ to itself that fixes three points. Therefore, that projectivity is the identity. So that composition maps any point $A$ of $EF$ to itself. So $A$, $A_2 = \phi_1 \circ \phi(A)$, and $B_2$ are collinear.                                □

**Corollary** *Given $\triangle DEF$ inscribed in the triangle with vertices $B$, $B_1$, and $B_2$, as pictured in Fig. 10.9. Let $A$ be a point of side (extended) $EF$. Then there is a triangle $AA_1A_2$ so $A_1$ lies on $DE$, $A_2$ lies on $DF$, with $B$ on $AA_1$, $B_1$ on $A_1A_2$, and $B_2$ on $AA_2$.*

Note 4  A pair of intersecting lines can be regarded as a conic. One way to see this as plausible is to note that the solution set of a second degree equation like $(2x - y)(3x - y + 2) = 0$ is a pair of lines. As there is a Cross-Joins Theorem for a hexagon inscribed in a pair of lines, so there is also a *Cross-Joins Theorem on a Conic*. Let $A, B, C, \ldots, A', B', C', \ldots$ be points of a conic and for some $X$ and $Y$ on the conic suppose the pencils $XA, XB, XC, \ldots$ and $YA', YB', YC', \ldots$ are projectively related. We say the points $\{A, B, C, \ldots\}$ and $\{A', B', C', \ldots\}$ are *projectively related sets of points on the given conic.*

**Theorem 10.10 (Cross-Joins Theorem on a Conic)** *Suppose $\{A, B, C, \ldots, P, Q, \ldots\}$ and $\{A', B', C', \ldots, P', Q' \ldots\}$ are projectively related sets of points on a given conic. Then the cross-joins $AB' \cap A'B$, $AC' \cap A'C$, $BC' \cap B'C$, $\ldots$ are collinear.*

**Proof** See Fig. 10.10. For points $P$ and $P'$, pencils $\{P'A, P'B, P'C, \ldots, P'P, \ldots\}$ and $\{PA', PB', PC', \ldots, PP', \ldots\}$ are projectively related. Since line $PP'$ corresponds to itself, then the corresponding lines of the pencils meet in collinear points on a line $m$: $P'A \cap PA'$, $P'B \cap PB'$, $P'C \cap PC'$, $\ldots$. Let $m$ meet the conic in point $V$. This means lines $PV$ and $P'V$ correspond, which means point $V$ corresponds to itself in the original correspondence of points of the conic. So if we take some other point $A$ of the conic and its corresponding $A'$, then these pencils $\{A'V, A'A, A'B, A'C, \ldots, A'P, \ldots\}$ and $\{AV, AA', AB', AC', \ldots, AP', \ldots\}$ are projectively related with $AA'$ related to itself. So their corresponding lines meet in collinear points, including $V$ and $AP' \cap A'P$, which must lie on $m$.                          □

**Fig. 10.10**  Cross-joins theorem on a conic

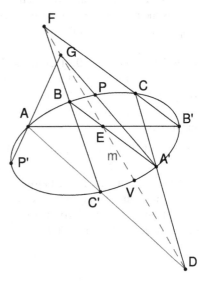

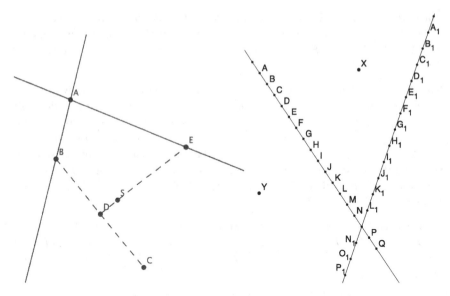

**Fig. 10.11** Left: Exercise 3. Right: Exercise 4

**Exercise 1** Referring to Fig. 10.1 of this chapter, describe a central collineation that produces the line-to-line perspectivity of Fig. 10.1 Left, and another central collineation that produces the pencil-to-pencil perspectivity of Fig. 10.1 Right.

**Exercise 2** Here is a special case of the Cross-Joins Theorem. Prove the following: Let lines $a$ and $a_1$ be projectively related, where they meet in point $E$ that is related to itself. Then we know the two lines are in perspective from a point $V$, with points $A_1, B_1, C_1, \ldots$ on $a_1$ joined, respectively, to points $A, B, C, \ldots$ on $a$ by lines that lie on $V$. Take the central collineation with center $V$, where the line on $E$ and $AB_1 \cap A_1B$ is the axis, and $A$ is mapped to $A_1$. Prove that (1) $B_1$ is mapped to $B$, (2) $B$ is mapped to $B_1$, (3) $AC_1 \cap A_1C$ lies on the axis.

**Exercise 3** See Fig. 10.11 Left. Use Maclaurin's construction to find points on the conic on $A, B, C, S, E$. Lines $AB$ and $AE$ are to be the two given lines, and $D = BC \cap ES$. We start with some line on $D$, meeting $AB$ at $Q$ and $AE$ at $R$; then $QS \cap RC$ is the point found on the conic. Construct enough points on the conic to see the shape.

**Exercise 4** See Fig. 10.11 Right. The two lines pictured are projectively related, as indicated by the lettering. (One is the image of the other by a dilation-rotation-translation sequence, and all those steps are compositions of central collineations.) Form the conic which is the set of points of intersection of projectively related pencils on $X$ and $Y$. Two answers are possible, depending on whether the lines on $X$ are drawn to points on the line to its right or to its left. Which conic?

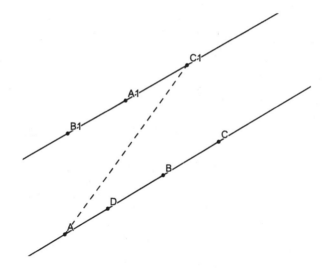

**Fig. 10.12** Exercise 5

**Exercise 5 (Proof of the Dual in Theorem 10.3)** Let there be projectively related pencils $\{a, b, c, \ldots\}$ and $\{a_1, b_1, c_1, \ldots\}$, on points $X$ and $X_1$, where the line $XX_1$ in one pencil is related to itself in the second pencil. Prove that the pencils are in perspective, i.e., corresponding lines meet on a line. (That line will be the line on $a \cap a_1$ and $b \cap b_1$.)

**Exercise 6** See Fig. 10.12. Map $A, B, C$ to $A_1, B_1, C_1$, respectively, by a composition of two perspectivities, first mapping line $AC$ to line $AC_1$, then mapping line $AC_1$ to line $A_1C_1$. Use that mapping to construct the image of $D$ by just a straightedge.

**Exercise 7** You can use Pappus's Theorem to prove the theorem of Chap. 2 Exercise 23. $a$. Let $\phi_A$ and $\phi_B$ be $(V, l)$ collineations with a common axis $l$ and a common center $V$, $V$ not on $l$. We have collinear $V, B, B'$ and collinear $V, A, A'$ where $V, A$ and $B$ are not collinear, with $\phi_B(B) = B'$ and $\phi_A(A) = A'$. Define $L$ to be $\phi_B(A')$ and $J$ to be $\phi_A(B)$. This means $AB' \cap A'J = I$ and $A'B \cap B'L = K$ lie on $l$. Let $AB$ meet $l$ at $H$. Prove $J, L, H$ are collinear. See Fig. 10.13. $b$. Now let $n$ be line $AB$. Prove $\phi_A \circ \phi_B(n)$ and $\phi_B \circ \phi_A(n)$ are the same line. This shows $\phi_A \circ \phi_B$ and $\phi_B \circ \phi_A$ are the same $(V, l)$ collineation.

**Exercise 8** From von Staudt [14, p 174]. Complete the proof.

**Theorem 10.11** *Let points $A, B, C, D$ lie on a circle (conic), and let $E$ be the polar of $AB$ while $F$ is the polar of $DC$. Then $A, B, C, D, E, F$ lie on a conic.*

**Fig. 10.13** Exercise 7

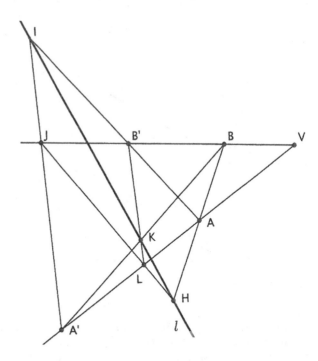

***Proof*** Let $J = AC \cap BD$, $K = AB \cap DC$, and $I = AD \cap BC$. *a*. Show $I$ is the polar of $JK$. *b*. Let $H$ be the polar of $BD$ and $G$ the polar of $AC$. Show $H, I, G$ are collinear.

*c*. Since $H, I, G$ are collinear, this means the opposite sides of hexagon $AEBCFD$ meet in collinear points. By the Converse of Pascal's Hexagon Theorem, Theorem 7.8, then hexagon $AEBCFD$ is inscribed in the unique conic that we know lies on $A, B, C, D, E$.                                                □

**Exercise 9** Here is the problem from Euclid's *Hyptios Porisms*, recorded in Pappus's *Mathematical Collection*, noted in Chap. 3. The wording is modernized, from [11, p 556]. See Fig. 10.14. The problem is to justify the claim.

> Let three variable straight lines $l_1, l_2, l_3$ meet straight line $l_4$ in, respectively, unvarying points $P_1, P_2, P_3$, while the intersection of $l_2$ and $l_3$ lies on a given straight line $m_2$ and the intersection of $l_1$ and $l_3$ lies on a given straight line $m_1$. Then it is possible to construct a straight line $m_3$ on which the intersection of $l_1$ and $l_2$ lies.

## 10.6   Some Hints and Solutions

**Exercise 3** Hint: Draw a line on $D$, meeting line $AB$ at $Q$ and line $AE$ at $R$. Then draw lines $QS$ and $CR$. Then $QS \cap CR$ will lie on the desired conic. Repeat for a different line on $D$. The conic will be a hyperbola.

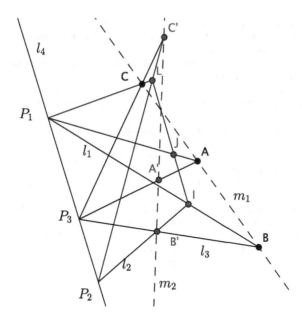

**Fig. 10.14** A Hyptios Porism, of Euclid, as reported in Pappus's *Mathematical Collection*, Book 7

**Exercise 4** Note: You will get either an ellipse or a hyperbola, ellipse if the line from $Y$ to $A$ is paired with the line on $X$ to $A_1$, etc.

**Exercise 7** Solution: For $a$., the vertices of hexagon $AA'IKB'B$ lie alternately on two lines. So the opposite sides of the hexagon meet in collinear point, namely, $L$, $J$, and $H$. For $b$., $\phi_A \circ \phi_B(B) = \phi_A(B') = J$ and $\phi_B \circ \phi_A(A) = \phi_B(A') = L$. $\phi_A \circ \phi_B$ and $\phi_B \circ \phi_A$ both fix point $H$, so each maps line $n$ to line $JL$.

**Exercise 8** Solution to $a$: $I$ is the polar of $JK$ since $I$, $J$, and $K$ are points where the diagonals of $ABCD$ meet. $b$. The polars of $H, I, G$ are concurrent at $J$.

**Exercise 9** Solution: Pencils on $P_1$ and $P_3$ are projectively related by meeting on $m_1$, with $l_4$ related to itself. Pencils on $P_2$ and $P_3$ are projectively related by meeting on $m_2$, with $l_4$ related to itself. So pencils on $P_1$ and $P_3$ are projectively related, with $l_4$ related to itself. So the corresponding lines meet in collinear points. (In the diagram, $P_2$, $A'$ and $J$ are collinear.)

# References

1. Charles Julien Brianchon, *Mémoire sur les Lignes du second Ordre*, Paris: Bachelier, 1817.
2. Michel Chasles, *Traité des Sections Coniques*, Paris: Gauthiers-Villars, 1865.
3. Julian Lowell Coolidge, *A History of Geometrical Methods*, Oxford University Press, 1940.

4. Luigi Cremona, *Elementi di Geometria Projettiva*, Rome: G. B. Paravia, 1873.
5. Luigi Cremona, *Elements of Projective Geometry* 2nd edition, transl. by C. Leudesdorf, Oxford: Clarenden Press, 1893.
6. Richard Courant and Herbert Robbins, *What is Mathematics?*, New York and London: Oxford University Press, 1941; reissued, revised by Ian Stewart, 1996.
7. L. Wayland Dowling, *Projective Geometry*, New York: McGraw-Hill, 1917.
8. J. D. Gergonne, Géométrie de situation, *Annales des Mathématiques*, vol 18, 1827–28, 149–154.
9. Colin Maclaurin, Letter from Mr. Colin Mac Laurin, in *Philosophical Transactions of the Royal Society*, Vol 39, Issue 439, 1735.
10. A. Moebius, *Der Barycentrische Calcul: ein neues Hülfsmittel zur analytischen Behandlung der Geometrie*, Leipzig: Barth, 1827.
11. Pappus of Alexandria: Book 7 of the *Collection*, edited by Alexander Jones. New York: Springer-Verlag, 1986.
12. J. V. Poncelet, *Traité des Propriétés Projectives des Figures*, Paris: Bachelier 1822.
13. Jacob Steiner, *Systematische Entwicklung der Abhägigkeit geometrischer Gestalten*, Erster Theil, Berlin: G. Fincke, 1832, in *Jacob Steiner's Gesammelte Werke*, Erster Band, edited by K. Weierstrass, Berlin: G. Reimer, 1881, 229–460.
14. K. G. C. von Staudt, *Geometrie der Lage*, Nürnberg: Bauer und Raspe, 1847; second edition 1856/1857.

# Chapter 11
# Desargues and Involution

## 11.1 Girard Desargues and Involution

Girard Desargues (1591–1661) is one of the most intriguing figures in the history of mathematics. He wrote a profound and bold booklet on conic sections, in 1639, which might have created projective geometry nearly two centuries before its actual birth. That work was *Brouillon project d'une atteinte aux événements des rencontres d'un cône avec un plan*, or *A Sketch of a Study of Sections of a Cone by a Plane*. Except the work was lost until La Hire's transcription was found about 1845. Even so, Poncelet knew about involution, an invention of Desargues, when he wrote his 1822 *Traité*; the *Brouillon project* had disappeared but letters written in the vitriolic controversy about the work had survived.

We know little about Desargues himself. He came from a wealthy Lyon family. In the 1630s he lived in Paris and became a member of Father Mersenne's circle of scientists who met regularly. He acquired a thorough knowledge of the Greek heritage in geometry. His works show an interest in applications of geometry, including a 12-page pamphlet on perspective drawing, *Exemple de l'une des manières universelles du S.G.D.L. touchant la pratique de la perspective*, of 1636, followed by a work on sundials and, with Abraham Bosse, a work on stone cutting. The theorem on perspective triangles which bears his name appeared in a work published by Bosse in 1648, but even that was little known. We know he was involved in several construction projects as architect, but it would be difficult to label his profession.

His 1639 work explicitly employed points at infinity, at which parallel lines are concurrent. Points in involution, and the invariance of that relation under projection, was a key concept. Corresponding to harmonic sets of collinear points, he introduced harmonic sets of concurrent lines. His work was projective in that properties of circles which are preserved under projection became properties of the conic sections. And he made thorough use of the pole-polar relation.

© Springer Nature Switzerland AG 2020
C. Baltus, *Collineations and Conic Sections*,
https://doi.org/10.1007/978-3-030-46287-1_11

**Fig. 11.1** Desargues'
Involution
Theorem—preliminary
version

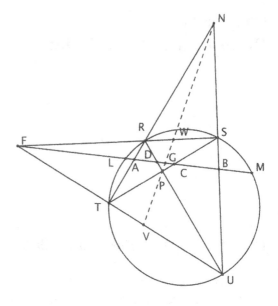

Here is a way of thinking, initially, of involution. See Fig. 11.1. On $F$ are secants meeting the given circle at $R$ and $S$, at $L$ and $M$, and at $T$ and $U$. Let $RT$ meet $SU$ at $N$, and $RU$ meet $ST$ at $P$. And $PN$ meets $TU$ at $V$ and $RS$ at $W$. As we have seen in the work of La Hire, $PN$ is the *polar* of $F$. Let $LM$ meet $RT$ at $A$, $SU$ at $B$, $RU$ at $D$, $ST$ at $C$, and $PN$ at $G$. Then, $H(FG, LM)$, $H(FV, TU)$, $H(FW, RS)$. By projection from $N$, $H(FG, AB)$. And by projection from $P$, $H(FG, CD)$. To arrive at Desargues' definition of points in involution, we recall that the relation $H(XY, ZW)$, which means $XW \cdot YZ = XZ \cdot YW$ with exactly one of $Z$ and $W$ between $X$ and $Y$, can also be written $QX^2 = QY^2 = QZ \cdot QW$ with the same requirement for order, when $Q$ is the midpoint of $XY$ (Theorem 4.4). Therefore, we have, when $Q$ is the midpoint of $FG$, $QG^2 = QL \cdot QM = QA \cdot QB = QC \cdot QD$.

**Definitions XXIV (Desargues)** When, on a line, for some point $Q$, which Desargues called the *souche* or *stump*, we have several pairs of distinct points so $QL \cdot QM = QA \cdot QB = QC \cdot QD = etc.$, then we say that those pairs $L, M$; $A, B$; $C, D$; *etc.* are *points in involution*, where $Q$ separates all or none of the pairs. Desargues called the collection of pairs an *arbre* ( tree).

Our discussion above proves this preliminary version of Desargues' Fundamental Involution Theorem in the case of a circle. See Fig. 11.1.

**Theorem 11.1** *Given Quadrilateral RSUT inscribed in a circle, $F = RS \cap TU$, with lines and points as in Fig. 11.1, then $L, M$; $A, B$; $D, C$ are pairs in involution.*

Desargues characterized points in involution by what we now see as an equality of cross-ratios:

**Theorem 11.2 (Desargues' Lemma)** *Let* $B, H$; $C, G$; $D, F$ *be three pairs in involution with* souche $A$. *Then*

$$\frac{GB \cdot GH}{CB \cdot CH} = \frac{GD \cdot GF}{CD \cdot CF} = \frac{AG}{AC}.$$

*(Corresponding equations hold for the other pairs.) [4, p 87] Conversely, when the first equality holds for collinear points* $B$, $H$, $C$, $G$, $D$, $F$, *then* $B, H$; $C, G$; $D, F$ *are three pairs in involution.*

***Proof*** By the definition of involution, $AB \cdot AH = AC \cdot AG = AD \cdot AF$ with $A$ separating all pairs or none. Since $\dfrac{AG}{AF} = \dfrac{AD}{AC}$ then by adding or subtracting in both the numerator and denominator, this fraction equals $\dfrac{GD}{FC}$. Likewise $\dfrac{AF}{AC} = \dfrac{AG}{AD}$ gives the equal fraction $\dfrac{GF}{CD}$. Thus, $\dfrac{GD \cdot GF}{FC \cdot CD} = \dfrac{AD}{AC} \cdot \dfrac{AG}{AD} = \dfrac{AG}{AC}$. In a similar way, $\dfrac{GB \cdot GH}{CB \cdot CH} = \dfrac{AG}{AC}$. (One can retrace steps in the proof to show the converse, that $\dfrac{GB \cdot GH}{CB \cdot CH} = \dfrac{GD \cdot GF}{CD \cdot CF}$ implies that the pairs $B, H$; $C, G$; $D, F$ are in involution with a *souche A*.)                                          □

**Corollary** *The involution relation is preserved under projection. Why? The cross-ratio is preserved under projection. (This was not Desargues' proof.)*

When Michel Chasles (1793–1880) published his *Traité de Géométrie Supérieure* in 1852, he could speak of points of a line projectively related to that same set of points, thanks to Steiner. Here, from Art. 182, is his definition of points in involution, which is equivalent to Desargues' Lemma.

> Let $a, a'$; $b, b'$ and $c, c'$ be three systems of two points, which are self-corresponding or *pairwise conjugate*, namely, $a$ and $a'$, $b$ and $b'$, $c$ and $c'$; if the cross-ratio of four of these, such as $a, b, c$ and $c'$, is equal to that of their conjugate points $a', b', c'$ and $c$, then these six points are said to be *in involution*.

It was Desargues who introduced the notion of involution, but the concept has expanded since his time. Involution for Desargues was a pairing of points. We now think of involution as a particular type of function that maps, projectively, the points of a line onto that same set of points. Such a function $f$ is an involution if whenever $f(x) = y$ then $f(y) = x$.

An important property is that a projective function on a line that exchanges two points, must necessarily exchange all points in pairs. See, for example, [8, p 54].

**Theorem 11.3** *Suppose the points of a line are projectively related to points of the same line, where* $X'$ *denotes the image of* $X$. *Suppose, further, that for some* $A$ *on the line* $A'' = A$, *i.e.,* $A'$ *is mapped to* $A$. *Then (1) for any* $X$ *on the line,* $X'' = X$ *and, (2), all the pairs* $A, A'$; $B, B'$; $C', C'$, *etc. are in involution.*

***Proof*** First we show $X'' = X$. Since the points are projectively related, the cross-ratio is preserved. So $CR(AA', XX') = CR(A'A, X'X'')$. But $CR(A'A, X'X'') = CR(AA', X''X')$, so $X$ must equal $X''$. Second we need to show that we have an involution as characterized by Desargues' Lemma, i.e., $CR(AA', BC') = CR(AA', CB')$ for all $A, B, C$. Since the cross ratio is preserved under projection, $CR(AA', BC) = CR(A'A, B'C')$, which means

$$\frac{AB \cdot A'C}{AC \cdot A'B} = \frac{A'B' \cdot AC'}{A'C' \cdot AB'}.$$

This equation shows $CR(AA', BC') = CR(AA', CB')$ since rearrangement gives

$$\frac{AB \cdot A'C'}{A'B \cdot AC'} = \frac{AC \cdot A'B'}{AB' \cdot A'C}. \qquad\qquad \square$$

To reach his Fundamental Involution Theorem, Desargues made repeated use of Menelaus's Theorem, as in his proof of the above corollary. He called it Ptolemy's Theorem, since Ptolemy had proved it in his widely known *Almagest*.

Here is Menelaus's Theorem: If collinear points $D$, $H$, $G$ lie, respectively, on sides (extended) $4h$, $hK$, and $K4$ of triangle $Kh4$, then

$$\frac{Dh}{D4} = \frac{Hh}{HK} \cdot \frac{GK}{G4}.$$

The proof is by triangle similarity in Fig. 11.2, where $KF$ is drawn parallel to line $HDG$.

Using Desargues' Lemma and Menelaus's Theorem, and the Power-of-a-Point Theorem, Desargues proved his Fundamental Involution Theorem in the case of a circle. Since the involution relation is preserved by projection, the theorem holds for a conic. Our proof uses concepts introduced by Steiner. The first appearance of the theorem after the seventeenth century may be in George Salmon's [6, p 279] of 1855, the third edition of an 1848 book. (I have not been able to locate a copy of the first

**Fig. 11.2** Menelaus's
Theorem

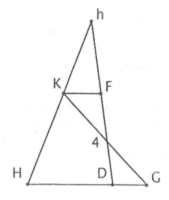

edition.) When Chasles developed the topic in [1], he attributed the development to Salmon. That Salmon did not credit the theorem to Desargues may indicate that he was not aware of the content of Desargues' [2], but had learned of involution.

**Theorem 11.4 (Desargues' Fundamental Involution Theorem)** *Given four points on a conic, R, S, T, and U, where RS and TU meet in F, RT meets SU at N, and RU meets ST at P. Suppose a line meets TU in Q and RS in W, meets the conic in L and M, meets opposite sides of quadrilateral RSUT in A and B , meets the diagonals in C and D, and meets NP in G. Then W, G; L, M; A, B; C, D are pairs of points in involution. The same relation holds for any conic on R, S, T, U (Fig. 11.3).*

***Proof*** [Based on [6] and [3]] See Figure 3, based on [7, p 142]. We prove the theorem for a circle, as had Desargues but not Salmon. Since involution is preserved by projection, the theorem will follow for a conic section. We consider two projectively related pencils. One, on $U$, includes lines $UL, UR, US, UM$, while the second, on $T$, includes the corresponding lines $TL, TR, TS, TM$. By the Inscribed Angle Theorem, corresponding angles between lines of the two pencils are equal, so the pencils are projectively related. So the points in which these lines meet line $LM$ are in a projective relation on $LM$, and the corresponding points are in equal cross-ratios: $CR(LM, DB) = CR(LM, AC) = CR(ML, CA)$. We define a projective relation of line $LM$ onto itself in which points $L$ and $M$ are interchanged and $D$ is mapped to $C$. That projective relation must be an involution, and by the equality of cross ratios, $B$ mapped to $A$. To involve $W$ and $Q$, consider pencils on $R$ and $U$. This time we pair lines $UL, UM, UT, US$ with, respectively, $RL, RM, RT, RS$. So $CR(LM, QB) = CR(LM, AW) = CR(ML, WA)$. The involution that interchanges $M$ and $L$ and maps $B$ to $A$, the same involution we created above, must interchange $W$ and $Q$. We conclude that pairs $L, M; A, B; C, D; W, Q$ are in involution.                                                    □

**Fig. 11.3** Desargues'
Fundamental Involution
Theorem

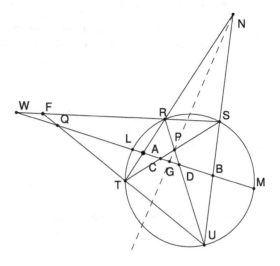

## 11.2   Foci in Desargues' Work

Now we look briefly at the treatment of foci in Desargues' 1639 work. He began with, essentially, the definition used in the *Conics* of Apollonius, for the ellipse and hyperbola, but then he characterized the foci in terms of involution. Following the interpretation of Desargues by Hogendijk [5], we have this theorem for an ellipse. See Fig. 11.4 Left.

**Theorem 11.5** *Given an ellipse with center $F$ and foci $Q$ and $X$, we pair a point $Z$, on the axis extended outside the ellipse, with the point $R$, on the axis inside the ellipse, by drawing a tangent $ZG$ and then, on $G$ on the ellipse, drawing the perpendicular to the tangent. $R$ is the point at which the perpendicular meets the axis. Then*

a. $H(QX, ZR)$,
b. *the resulting pairs $Z_1, R_1$; $Z_2, R_2$; $Z_3, R_3$; ... are points in involution with souche $F$, and when the involution is extended to the other points of the axis, $Q$ and $X$ are fixed points of the involution.*

***Proof*** It follows from a conic property, Theorem 9.3*a*, that the perpendicular $GR$ to the tangent at $G$ bisects $\angle QGX$. By a converse of the lemma in the proof of Theorem 9.5, this means that lines $GZ, GQ, GR, GX$ form a harmonic pencil. Therefore, $H(QX, ZR)$. Then by Theorem 4.4, $H(QX, ZR)$ is equivalent to $FX^2 = FQ^2 = FR \cdot FZ$, when $F$ is the midpoint of $QX$. Therefore, $FR_1 \cdot FZ_1 = FR_2 \cdot FZ_2 = FR_3 \cdot FZ_3 = \ldots$ . This means $Z_1, R_1$; $Z_2, R_2$; $Z_3, R_3$; ... are points in involution with *souche $F$*. Finally—and we do not find this in Desargues' work—points $Z$ and $R$ are paired not just by the tangent construction, but also $Z$ and $R$ are paired when $H(QX, ZR)$. Now, the pairing of $Z$ and $R$ is projective since it can be carried out by a central collineation, and a central collineation is projective. So we have an involution of the points of the axis of the conic, and $Q$ and $X$ are the fixed points of that involution.          □

The fixed points, $Q$ and $X$, of the involution described in the above theorem, are the *foci* of the ellipse. They were labeled *double nodes* or *noeuds moyen double* by Desargues.

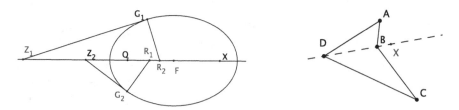

**Fig. 11.4** Left: Foci in Desargues, based on Hogendijk. Right: Exercise 1

Hogendijk[5] makes a compelling case that much of the *Brouillon Project*, especially parts most puzzling to readers, was to reach the claim that the foci are the double points, *noeuds moyen double*, of a particular involution. Theorem 11.5 is a part of that story.

## 11.3   Notes and Exercises

**Note 1**   An involution of a given line that has two fixed points can be constructed by the central collineation that we have called the *involution collineation*. See Theorem 7.2. (An involution need have no fixed points; it cannot have exactly one [8, p 56].) One fixed point will be the center of the central collineation, while the other is on the axis.

**Exercise 1**   This problem is really about involution.

*a.* Given quadrilateral $ABCD$, find the center $V$ of a central collineation for which $DB$ is the axis and $A$ is mapped to $C$, and $C$ is mapped to $A$. See Fig. 11.4 Right. Hint: Let $X = AC \cap BD$. $V$ must lie on line $AC$. The result will be more familiar if $ABCD$ is not convex, say with $D - B - X$. Let $AD$ meet $BC$ at $S$ and let $CD$ meet $AB$ at $R$. Now find $V$ and explain why the central collineation exchanges $A$ and $C$.

*b.* Now explain, based on your diagram for part *a.*, why $A$ and $C$ are harmonic conjugates of $X$ and $V$. (This is a further hint for the construction asked for in part *a.*)

## References

1. Michel Chasles, *Traité des Sections Coniques*, Paris: Gauthiers-Villars, 1865.
2. Girard Desargues, *Brouillon project d'une atteinte aux événements des rencontres d'un cône avec un plan*, in [4] (trans Field) and original http://gallica.bnf.fr/ark:/12148/bpt6k105071b/f1. image Paris, 1639.
3. Heinrich Dörrie, *100 Great Problems of Elementary Mathematics*, New York: Dover, 1965, from the Fifth Edition in German, 1958.
4. J. V. Field and J. J. Gray, *The Geometrical Work of Girard Desargues*, New York: Springer, 1987.
5. Jan P. Hogendijk, Desargues' *Brouillon Project* and the *Conics* of Apollonius, *Centaurus* 34, 1991, 1–43.
6. George Salmon, *A Treatise of Conic Sections*, Third Edition, London: Longman, Brown, Green, and Longmans, 1855.
7. René Taton, *L'oeuvre mathématique de G. Desargues*, Paris: Presses Universitaire de France, 1951.
8. John Wesley Young, *Projective Geometry*, Mathematical Association of America, Chicago: Open Court Publishing, 1930.

# Chapter 12
# Looking Ahead

## 12.1 Projective Geometry After Steiner

This introductory treatment of projective geometry essentially ends with Jacob Steiner's work. But it is good to see several of the changes, after Steiner's work, which have been central in the development of the subject since Steiner's time.

First, two prominent geometers continued along the path set by Poncelet. These are Michel Chasles (1793–1880) and Luigi Cremona (1830–1903). Chasles was a near contemporary of Poncelet, entering the *École Polytechnique* in 1812, but his career carried on long past the retirement of Poncelet from work in projective geometry. He placed the invariance of the cross-ratio, which he called *anharmonic*, under projection, as a foundation of his geometry. We have seen that he developed properties of involution, adopted a definition of foci that was suggested by Poncelet, and made an important repair to Steiner's treatment of conics. He wrote an important history of geometry [2], that is still worth studying, and wrote major texts in 1852 [3] and 1865 [4] that helped fix projective geometry as a staple of university education in mathematics.

Luigi Cremona published *Elementi di geometria projettiva* in 1873 [7]. He was already a leading researcher in curves and surfaces, including a prize-winning memoire in 1866 on surfaces of degree three. His 1873 work was the first to place the elaboration of projection by straightedge construction at the beginning, as a core concept. What we have called a central collineation, he called a *homology*. After introducing homology, Cremona gave several striking diagrams of point-by-point straightedge constructions of conics, in the manner described earlier. (See, for example, *Cremona* 1893 *Fig.* 10, in our Fig. 12.1.) Cremona had hoped his book would serve in secondary schools; this dream was not realized. His work did find favor among university students and professors, and Cremona collaborated with translators to bring out French, German, and English editions. But his work was the end of an era rather than a beginning.

© Springer Nature Switzerland AG 2020

C. Baltus, *Collineations and Conic Sections*,
https://doi.org/10.1007/978-3-030-46287-1_12

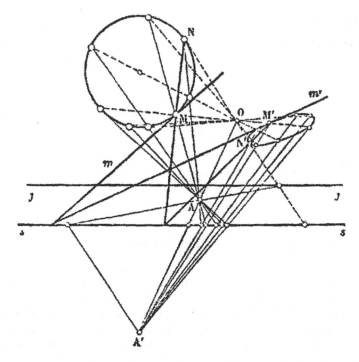

**Fig. 12.1** Cremona, *Fig.* 10, 1893. Central Collineation of a circle. $O$ is the center, $s$, the lower horizontal line, is the axis, point $A$ is mapped to $A'$, and $j$ is the vanishing line

We have already seen one instance in which a concept defined by length, the harmonic relation, was reworked by Carnot to depend on the complete quadrilateral, which does not require the concept of length. The cross-ratio, based on length, which was a central concept for Steiner and Chasles, is often relegated to the back of newer books, as with Young and Pedoe, or not mentioned at all, as with Coxeter [6]. K. G. C. von Staudt [12, 1847] led the movement away from metrical considerations. Where Steiner and Chasles made the preservation of the cross-ratio among collinear points the defining characteristic of projectively related forms, Von Staudt returned to the harmonic relation, which can be defined without length in terms of a complete quadrilateral. Two forms are projectively related when any four harmonic elements of one correspond to harmonic elements of the other. But he showed a way, in [12] and [13], to allow for arithmetic operations on collinear points. Young [16, p 120] spoke of this achievement as

> the introduction of analytic methods on a purely projective basis, that is, without the use of any metric notions whatever.

A second change in projective geometry was the attention to foundations: what do we need to assume about a projective plane to arrive at the conclusions we find important? In affine geometry, that based on Euclid's *Elements*, various revisions to

Euclid's postulates were suggested in the late 1800s, culminating in David Hilbert's *Grundlagen der Geometrie*, 1899. There was corresponding work at the same time in projective geometry, culminating in Pieri's *Principles of the Geometry of Position Composed in a Logical-Deductive System*, of 1899 [11]. It had 19 axioms. Coxeter's *Introduction to Geometry* presents projective geometry based on six axioms [6, p 230], based on work of F. Bachmann. It is notable that Pieri had published an edited translation of G.K.C. von Staudt's *Geometrie der Lage*; his work on foundations of projective geometry was, in large part, to repair deficiencies of Von Staudt's treatment. See [10], which includes a nice exposition of Von Staudt's geometry.

In all the proposed axioms for projective geometry, we see one or several axioms which requires any pair of lines to meet. Veblen and Young, in 1910, had Assumption III: "Any two m-classes [lines] have at least one element of $S$ [set of points] in common." [14] Pieri, on the other hand, had several axioms from which this same result can be deduced.

At this point, one could simply take a projective plane as the set of "points," with particular subsets designated as "lines," satisfying a particular set of axioms. Then operations on those points could be defined.

With the examination of foundations came a widespread conviction that projective geometry was more basic than classical Euclidean geometry, " 'prior to' rather than 'derivative from' Euclidean geometry." [10, p 285] This reversal of positions became an area of study in itself, going back to, at least, Arthur Cayley, in the 1850s. See [8, p 239–271]. Where Euclid's *Elements*, even without the Parallel Postulate, in Propositions 16 and 27 of Book 1, made parallel lines necessary, and the line at infinity was only an artificial addition to the plane, now the tables were turned. One would have to remove a line from the primordial projective plane to get the affine geometry that Euclid created.

## 12.2 Notes

1. The fate of La Hire's work on central collineations is worth noting. Chasles's *Aperçu historique sur l'origine et developpement des méthodes en géométrie* (1837), apparently the first to recognize La Hire's work in the *Plani-coniques*, also noted that [9] was "extremely rare" [2, Article 30]. Cremona did not mention La Hire's 1673 work in his historical summary in 1873; it is possible he did not know of the *Plani-coniques* or he may simply have recognized that, by 1873, La Hire's work had nothing new to contribute to projective geometry. Important historical works of Moritz Cantor and Julian Coolidge are exceptions that prove the rule. Cantor devoted a paragraph to the *Plani-coniques*, describing the construction process, but suggesting a crucial proof was complicated, and that he never saw the work [1, p 125]. Coolidge wrote

   > Another writer of this period [1600s] who made use of central projection was La Hire. His first venture was in 1673 when he wrote on 'plano-conics'. This seems to have been a failure. [5, p 90]

but he does not include La Hire's 1673 work in his bibliography. He, too, may never have seen the work. There is one full article devoted to the *Plani-coniques*, [15]. The author, H. Wieleitner, gave a careful exposition and tracked down known copies in 1913: he found just four copies of [9] which included the *Plani-coniques*, which may explain the failure of Cantor and Coolidge to read it. Fortunately, [9], with *Plani-coniques*, is now available, although there are missing diagram plates, at *gallica.bnf.fr*.

2. Determining a set of axioms for projective geometry has proven to be a messy business. A reader who searches among texts for a definitive set of axioms will be disappointed. Beyond our Properties $P1.$ and $P2.$ and some version of $P3.$, proposed collections differ substantially. A thoughtful historical review of the subject is in the later chapters of Jeremy Gray's [8].

# References

1. M. Cantor, *Vorlesungen über Geschichte der Matematik*, Dritter Band, Stuttgart: Teubner, 1901; reprinted New York: Johnson Reprint Corp., 1965.
2. Michel Chasles, *Aperçu historique sur l'origine et developpement des méthodes en géométrie*, Brussels, 1837.
3. Michel Chasles, *Traité de Géométrie Supérieure*, Paris: Bachelier, 1852.
4. Michel Chasles, *Traité des Sections Coniques*, Paris: Gauthiers-Villars, 1865.
5. Julian Lowell Coolidge, *A History of Geometrical Methods*, Oxford University Press, 1940.
6. H. S. M. Coxeter, *Introduction to Geometry*, New York: Wiley, 1969.
7. Luigi Cremona, *Elementi di Geometria Projettiva*, Rome: G. B. Paravia, 1873.
8. Jeremy Gray, *Worlds Out of Nothing: A Course in the History of Geometry in the 19th Century*, London: Springer-Verlag, 2007.
9. Philippe de La Hire, *Nouvelle Méthode en Géométrie pour les Sections des Superficies coniques et Cylindriques*, Paris, 1673.
10. Elena Marchisotto, The projective geometry of Mario Pieri: A legacy of Georg Karl Christian von Staudt, *Historia Mathematica* 33 No. 3, 2006, 277–314.
11. Mario Pieri, I principii della geometria di posizione composti in sistema logico deductivo, *Memoria della Reale Accademia della Scienze di Torino* 48 (2), 1899, 1–62.
12. K. G. C. von Staudt, *Geometrie der Lage*, Nürnberg: Bauer und Raspe, 1847; second edition 1856/1857.
13. K. G. C. von Staudt, *Beiträge zur Geometrie der Lage*, Nürnberg: Bauer und Raspe, 1856–1860.
14. Oswald Veblen and John Wesley Young, *Projective Geometry*, Boston, MA: Ginn, 1910.
15. H. Wieleitner, Über die 'Plani-coniques' von de La Hire, *Archiv für die Geschichte der Naturalwissenschaften und der Technik* Vol. 5, 1913 Leipzig, 49–55.
16. John Wesley Young, *Projective Geometry*, Mathematical Association of America, Chicago: Open Court Publishing, 1930.

# Chapter 13
# Matrices and Homogeneous Coordinates

## 13.1 Matrices for Collineations

As shown in Chap. 1, we can define the real projective plane as a set of ordered triples in a way that points at infinity are included. Further, a collineation $\phi$ can be defined by matrix multiplication: $\phi(\vec{x}) = M\vec{x}$ where $M$ is a 3-by-3 non-singular matrix and $\vec{x}$ is an ordered triple. (A 3-by-3 matrix is *non-singular* when its determinant is not zero.)

Such ordered triples first appeared in an 1827 work by A. Moebius. There, the triples were barycentric coordinates, a concept born in the physical concept of a center of mass, although these are not generally used in modern projective geometry.

Here is the idea in a specific example. Let $A = (1, 1)$, $B = (0, 2)$, $C = (0, 1)$. If we imagine a plate of weightless material except for weights $r$, $s$, $t$ placed at the three points $A$, $B$, $C$, then the center of mass, or centroid, of the plate is $[rA + sB + tC]/(r + s + t)$. If $r = 1$, $s = 2$, and $t = 3$, then the center of mass is $[(1, 1) + 2(0, 2) + 3(0, 1)]/6 = (1/6, 8/6)$. So the barycentric coordinates of $(1/6, 4/3)$ are $(1, 2, 3)$. We note that the same point $(1/6, 4/3)$ would also have barycentric coordinates $(3, 6, 9)$.

This last observation refers to the *homogeneity property*: if $k \neq 0$, then $(r, s, t)$ and $(kr, ks, kt)$ represent the same point.

Also note that coordinates can be negative, in which case the point represented by $(r, s, t)$, when $r$, $s$, $t$ are not all negative, will lie outside $\triangle ABC$. So $(2, -2, 1)$ represents point $[2(1, 1) - 2(0, 2) + (0, 1)]/1 = (2, -1)$.

And, we need $r + s + t \neq 0$. Consider a point where $r + s + t$ is nearly 0. With $A$, $B$, $C$ as above, consider $(r, s, t) = (-3, 2, b)$. Then

$$[rA + sB + tC]/(r + s + t) = (-3, 1 + b)/(b - 1).$$

© Springer Nature Switzerland AG 2020
C. Baltus, *Collineations and Conic Sections*,
https://doi.org/10.1007/978-3-030-46287-1_13

If $b = 1.1$, the point represented is (-30, 21). And as $b \rightarrow 1^+$, then $(-3, 1 + b)/(b - 1)$ approaches $\infty$ along the line $y = -\frac{2}{3}x$. In this sense we could say that $(-3, 2, 1)$ represents $\infty$ on the line $y = -\frac{2}{3}x$.

Now, let us move away from this musing to the prevailing system of homogeneous coordinates, which we defined as constituting the real projective plane, $\mathbf{P^2}$, in Chap. 1. Homogeneous coordinates first appeared in Julius Plücker's *Analytisch-Geometrische Entwicklungen, Methods of Analytic Geometry*, of 1828 [4], although there is some question as to whether his use was systematic [3].

As a reminder,

The *real projective plane*, which we denote $\mathbf{P^2}$, is the set of points $\{(x, y, z) :  x,  y,  z \text{ are real and not all } 0\}$, where the lines are the triples $(a, b, c)$ with $a, b, c$ real and not all 0. Point $(x, y, z)$ lies on line $(a, b, c)$ exactly when $ax + by + cz = 0$. The homogeneity property holds: for any non-zero $k$, triples $(x, y, z)$ and $(kx, ky, kz)$ represent the same point, while $(a, b, c)$ and $(ka, kb, kc)$ represent the same line.

*Vector* and *dot product* notation will help. Triple $(x, y, z)$ can be represented by a vector, such as $\vec{x}$. When triple $(a, b, c)$ is represented by $\vec{m}$, then dot product $\vec{x} \cdot \vec{m}$ denotes expression $ax + by + cz$.

As shown in Chap. 1, Properties $P1$. and $P2$. of projective geometry are satisfied: on any two points there is exactly one line and any two lines lie on exactly one point.

We assume the following, where $M$ represents a 3-by-3 matrix.

**Theorem 13.1**

1. *Point $\vec{x}$ lies on line $\vec{m}$ exactly when the dot product $\vec{x} \cdot \vec{m}$ equals 0.*
2. *Matrix $M$ is non-singular iff $M$ is invertible iff equation $M\vec{x} = \vec{b}$ has exactly one solution $\vec{x}$ for each constant $\vec{b}$.*
3. *Since for scalar $k \neq 0$, $M\vec{x} = \vec{b}$ exactly when $Mk\vec{x} = k\vec{b}$ then the function $\phi$ operating on the projective plane (in homogeneous coordinates) defined by $\phi(\vec{x}) = M\vec{x}$ is well defined. When $M$ is non-singular, then the function is one-to-one.*
4. *Let $M$ be a singular 3-by-3 matrix. Then there is a non-zero vector $\vec{a}$ such that $M\vec{a} = \vec{0}$.*
5. *When are three given points collinear?*
   *Points $\vec{x}_1, \vec{x}_2, \vec{x}_3$ are collinear exactly when some line $\vec{m}$ gives $X\vec{m} = \vec{0}$ where $X$ is the matrix whose rows are $\vec{x}_1, \vec{x}_2, \vec{x}_3$. There is such a non-zero vector $\vec{m}$ exactly when $det(X) = 0$.* $\square$

We give three important properties in the following theorems.

**Theorem 13.2** *If a function $\phi$ is defined on the projective plane by $\phi(\vec{x}) = M\vec{x}$ for a non-singular matrix $M$, then $\phi$ is a collineation.*

**Proof** When $x_1, x_2, x_3$ are three collinear points, then the 3-by-3 matrix $X = [\vec{x}_1 \ \vec{x}_2 \ \vec{x}_3]$ has determinant 0. The images of these points under $\phi$

are $M\vec{x}_1$, $M\vec{x}_2$, $M\vec{x}_3$, the columns of matrix $MX$. Since $det(X) = 0$, then $det(MX) = 0$, so the three image vectors are collinear. $\qquad\square$

**Theorem 13.3 (Moebius, 1827)** *If $a_1$, $a_2$, $a_3$, $a_4$ and $b_1$, $b_2$, $b_3$, $b_4$ are each sets of four points of the projective plane where no three are collinear, then there is a matrix M so for $i = 1$, 2, 3, 4, $M\vec{a}_i = \vec{b}_i$.*

**Proof** There are non-zero scalars $k_1, k_2, k_3$ and $j_1, j_2, j_3$ so $a_4 = k_1 a_1 + k_2 a_2 + k_3 a_3$ and $b_4 = j_1 b_1 + j_2 b_2 + j_3 b_3$.
We can find matrix $M$ so

$$M\vec{a}_n = \frac{j_n}{k_n}\,\vec{b}_n \quad \text{for } n = 1, 2, 3.$$

Then

$$M\vec{a}_4 = M[k_1 a_1 + k_2 a_2 + k_3 a_3] = k_1 M a_1 + k_2 M a_2 + k_3 M a_3$$

$$= k_1\frac{j_1}{k_1}b_1 + k_2\frac{j_2}{k_2}b_2 + k_3\frac{j_3}{k_3}b_3 = \vec{b}_4. \qquad\square$$

Moebius proved this theorem for a collineation, although he did not use matrices [2, Part 2 Art. 220].

**Theorem 13.4** *If function $\phi$ is defined on the projective plane by $\phi(\vec{x}) = M\vec{x}$ for a non-singular matrix M, then $\phi$ has a fixed point, $\vec{v}$.*

**Proof** We show there is a non-zero vector $\vec{v}$ so for some real $k$, $M\vec{v} = k\vec{v}$. ($\vec{v}$ is an *eigenvector* and $k$ is the corresponding *eigenvalue*.) Equation $M\vec{v} = k\vec{v}$ is equivalent to $(M - kI)\vec{v} = \vec{0}$. This equation has a non-zero solution $\vec{v}$ exactly when $det(M - kI)$ is 0. When $det(M - kI)$ is expanded by the rules for computing determinants, it is a cubic polynomial in $k$. A cubic polynomial with real coefficients has at least one real root, $k$. So for that value of $k$ there is a non-zero *eigenvector*. Because we deal with homogeneous coordinates, when $M\vec{v} = k\vec{v}$ for a triple $\vec{v}$, then, as a point in the projective plane, $\vec{v}$ is a fixed point. $\qquad\square$

We observe that not all collineations are central collineations. An example would be a rotation, that is not a half-turn, about a point. This suggest an important question:

When is a collineation, defined by matrix multiplication in the projective plane, a central collineation?

We know a collineation is a central collineation exactly when it has a line of fixed points. Let us consider the line on two given points, $\vec{a}_1$ and $\vec{a}_2$. This line is the set

$$\{t\vec{a}_1 + (1 - t)\vec{a}_2 : t \in \mathcal{R}\}.$$

Suppose, first, that $\vec{a}_1$ and $\vec{a}_2$ are independent eigenvectors with the same eigenvalue, $k$. (This means they do not represent the same point in the projective plane.) Then for a particular $t$,

$$M(t\vec{a}_1 + (1 - t)\vec{a}_2) = tM\vec{a}_1 + (1 - t)M\vec{a}_2 = k(t\vec{a}_1 + (1 - t)\vec{a}_2).$$

This means $t\vec{a}_1 + (1 - t)\vec{a}_2$ is a fixed point, so the line is a line of fixed points.

On the other hand, only eigenvectors are fixed points, so the only other possibility for a line of fixed points would be a line on two eigenvectors, $\vec{a}_1$ and $\vec{a}_2$, which have different eigenvalues, $k_1$ and $k_2$. Then

$$M(t\vec{a}_1 + (1 - t)\vec{a}_2) = tM\vec{a}_1 + (1 - t)M\vec{a}_2 = tk_1\vec{a}_1 + (1 - t)k_2\vec{a}_2.$$

Since $k_1 \neq k_2$, we do not have a fixed point.

We summarize

**Theorem 13.5** *A collineation of the projective plane, defined by matrix multiplication, is a central collineation exactly when there are at least two independent eigenvectors (triples) for one eigenvalue.*

## 13.2   Excursion: A 13-Point Projective Plane, and Yet Another Definition of a Conic

At the end of Chap. 1, we introduced a 7-point projective plane satisfying properties $P1.$, $P2.$, and $P3$. The number of meaningful claims about this projective plane is very limited. Just for starters, Desargues' Theorem makes no sense here. However, if, following the last problem at the end of the same chapter, we form a 13-point projective plane, then we have a far richer projective geometry.

Starting late in the nineteenth century, a geometry could simply be introduced as a set of points, with specified subsets designated as lines, satisfying a given set of axioms. Here we give an example of a 13-point geometry satisfying the set of five axioms set out by H. S. M. Coxeter on p. 95 of *Projective Geometry* [1]. In addition to Coxeter's first three axioms, which are properties $P1.$, $P2.$, and $P3.$ of our Chap. 1, Coxeter's fourth axiom requires that for any complete quadrangle, the three diagonal points not be collinear.

For the 13-point geometry, we take the affine plane of ordered pairs of numbers in the field $F_3$. $F_3$ consists of numbers 0, 1, 2 under addition and multiplication (*mod* 3). The nine points of the affine geometry are represented by homogeneous coordinates $(x, y, z)$, with $x$ and $y$ in $F_3$ and $z = 1$. We add in four new *points at infinity* by allowing $z$ to be 0, but not permitting a point $(0, 0, 0)$. Lines of the affine plane are defined as the solution sets of equations $ax + by + cz = 0$, with $a$ and $b$ not both 0. Each line is represented by the homogeneous triple $[a, b, c]$. The four points at infinity form a *line at infinity*, whose coordinates are [0, 0, 1]. The brackets will alert us that a particular triple is a line and not a point.

We note that the matrix Theorems 13.1 through 13.3 all hold. A non-singular matrix is one whose determinant is not 0 (*mod* 3). Since $F_3$ is a field, the derivations of these theorems hold as they do for real number matrices. It makes sense to speak of a complete quadrangle, so a harmonic set can be defined. A projectivity of the plane will be a mapping defined by multiplication by a non-singular matrix.

a. List the 13 elements of this projective plane as ordered triples, and list the 13 lines, again as ordered triples. How many points are on each line?

b. Take the four points, where no three are collinear, $A = (1, 0, 1)$, $B = (1, 1, 1)$, $C = (2, 0, 1)$, $D = (2, 1, 1)$, and find the three diagonal points: $AB \cap CD$, $AC \cap BD$, $AD \cap BC$, and show that those three points are not collinear.

c. Theorem 13.3, that any four given points, no three collinear, can be mapped by a collineation to any four given points, no three collinear, still applies. So the four points of some given quadrangle, no three collinear, are images of the points $A, B, C, D$ under a collineation. This means that the three diagonal points for the given quadrangle are not collinear. This geometry, therefore, satisfies Coxeter's fourth axiom.

The fifth axiom, that any projectivity of a line to itself which leaves three points unchanged must be the identity, holds in a trivial sense since each line has just four points.

Now, as for another definition of a conic, we need the idea of a *polarity*. In a plane, a *polarity* is a one-to-one pairing of points with lines, and of lines with points, that is of order 2. This means that if a point $X$ is paired with a line $x$, called the *polar* of $X$, then, in turn, $x$ is paired with $X$, its *pole*. A point $X$ is called *self-conjugate* if $X$ lies on its polar. We have encountered the pairing of points and lines of the projective plane defined by a given conic section: each point is paired with its polar as determined by that conic, and each line, likewise, is paired with is pole. The self-conjugate points with respect to a conic, under this polarity, are the points of the conic, where each point is paired with the tangent at that point. However, as we will see in a simple example, the polarity can be defined in other ways, and in such a case, the self-conjugate points form, by definition, a conic section.

The first mathematician to develop the concept of a polarity, or *Polarsystem*, and of a corresponding conic, was von Staudt, in [5] of 1847. He does follow Steiner's characterization of a conic, which he calls a *Curve II. Ordnung*, as the set of points of intersection of two projectively related pencils [p 173]. However, he also tells us [p 137]

In a planar *Polarsystem*, either no point is self-conjugate or there is a curve of even order ... so every point which belongs to the curve lies on its polar, which it attaches to itself, and so every point not on the curve, *der Ordnungscurve des Polarsystems*, lies outside its polar. No line meets the curve in more than two points, and there will be a tangent at every point [of the curve] and every other line on the point will cut the curve in exactly one other point.

*d. d.* Here is an obvious polarity for the 13-point geometry described above: pair each point $(x, y, z)$ with the line $[x, y, z]$ and vice versa. There are four self-conjugate points, points which lie on their polar. What are they? Then verify von Staudt's claim for one of those self-conjugate points, namely that of the four lines on the point, exactly one is tangent—meeting the conic in just one point—while each of the other three meets the conic in exactly one other point. Finally, give the second degree equation of the conic formed this way.

Answer to *d.*: $x^2 + y^2 = 2(mod\ 3)$.

## 13.3   Exercises

**Exercise 13.1** Suppose that $M$ is a non-singular 3-by-3 matrix. Let function $\phi$ be defined on the projective plane of points by matrix multiplication: $\phi(\vec{x}) = M\vec{x}$.

*a.* $\phi$ is also a function that maps each line to a line. When a line is represented by a triple of numbers $\vec{m}$, what is the matrix $L$ so $\phi(\vec{m}) = L\vec{m}$? (Describe $L$ in terms of $M$.)

*b.* Show that incidence is preserved, i.e., point $\vec{x}$ is on line $\vec{m}$ exactly when $M\vec{x}$ is on line $L\vec{m}$.

**Exercise 13.2** Explain: three lines $m_1$, $m_2$, $m_3$ are concurrent iff the matrix whose rows are $m_1$, $m_2$, $m_3$ is singular.

**Exercise 13.3** Suppose $M$ is a non-singular matrix with eigenvalue $k$ with exactly two independent eigenvectors $\vec{a}_1$ and $\vec{a}_2$. Prove there is another eigenvalue $j$ whose eigenvector, as a point in the projective plane, is not on the line on $\vec{a}_1$ and $\vec{a}_2$.

**Exercise 13.4** Now suppose $M$ is a non-singular matrix with eigenvalue $k$ with three independent eigenvectors: $\vec{a}_1$, $\vec{a}_2$, and $\vec{a}_3$. (So no one of $\vec{a}_1$, $\vec{a}_2$, and $\vec{a}_3$ is a linear combination of the other two.) Describe the corresponding central collineation of the projective plane. How do you know?

**Exercise 13.5** Given the collineation on the projective plane, in homogeneous coordinates, defined by matrix multiplication by matrix

$$M = \begin{pmatrix} 0 & 1 & -1 \\ 2 & 1 & 1 \\ 4 & -2 & 4 \end{pmatrix},$$

(1) How do you know the collineation is a central collineation?
(2) Write, in form $Ax + By + Cz = 0$, the equation of the axis. Then write the equation in form $y = mx + b$.
(3) Give the center of the central collineation, in form $(x, y, 1)$.
(4) Find the image of point $(-1, 0, 1)$.

(5) Give the equation of the vanishing line.
(6) When the circle $x^2 + y^2 = 1$, which is $x^2 + y^2 - z^2 = 0$ in homogeneous coordinates, is mapped by this central collineation, which conic is the image? Find the center of that conic. What else can you find?

**Exercise 13.6** This exercise leads to a coordinate proof of Pappus's Hexagon Theorem.

a. Let $\vec{p}_1, \vec{p}_2, \vec{p}_3$ be collinear points in the real projective plane. So the matrix whose columns are $\vec{p}_1, \vec{p}_2, \vec{p}_3$ is singular. Prove: In some representation in homogeneous coordinates $\vec{p}_1 + \vec{p}_2 = \vec{p}_3$, and for another representation, $\vec{p}_1 - \vec{p}_2 = \vec{p}_3$ . This means there are scalars $j$ and $k$ so that, as triples of real numbers, $j\vec{p}_1 \pm k\vec{p}_2 = \vec{p}_3$. (A "representation" of a vector $\vec{p}_1$ is a vector $k\vec{p}_1$ for some non-zero scalar $k$. )

b. Let $A = (1, 0, 0)$, $B = (0, 1, 0)$, $C = (0, 0, 1)$, $D = (1, 1, 1)$. Take a point collinear with $A$ and $B$, which we may designate as $E = A - B = (1, -1, 0)$, and a point collinear with $C$ and $D$, which we designate as $F = C + D = (1, 1, 2)$. To prove Pappus's Hexagon Theorem for hexagon $ACBDEF$, find the triples for the lines on opposite sides of the hexagon, then show that those opposite sides meet in collinear points. Explain why this proof by coordinates is, indeed, a proof of the Pappus Hexagon Theorem.

## 13.4  Some Hints and Solutions

### Exercise 13.1
*Solution to a.: First note that $L$ must exist and be a non-singular matrix since $\phi$ is a one-to-one mapping of lines to lines. For a given line $\vec{m}$, point $\vec{x}$ lies on $\vec{m}$ exactly when matrix product $\vec{x}^T \vec{m}$ is 0, which is equivalent to $\vec{x}^T L^{-1} L\vec{m} = 0$. Further, $\vec{x}^T \vec{m} = 0$ exactly when $(M\vec{x})^T$ is on line $L\vec{m}$, which is equivalent to $(M\vec{x})^T L\vec{m} = 0$. So for a fixed line $\vec{m}$, $\vec{x}^T L^{-1} L\vec{m} = 0$ and $(M\vec{x})^T L\vec{m} = 0$ are equivalent (for all $\vec{x}$), which is true exactly when $L^{-1} = M^T$.*

### Exercise 13.2
*Hint: See Part 5. of Theorem 13.1.*

### Exercise 13.3
*Solution: We have already seen in the discussion leading to Theorem 13.5 that the line on $\vec{a}_1$ and $\vec{a}_2$ is a line of fixed points in the projective plane. As shown in the first chapter, a collineation with a line of fixed points must have a center, $V$, which is a fixed point. $V$ cannot lie on the line of fixed points since in that case $V$ would be, as a triple, another eigenvector with eigenvalue $k$. So $V$, as a triple, is an eigenvector with a different eigenvalue.*

## Exercise 13.4

*Solution: The central collineation will be the identity. When $\vec{a}_1$, $\vec{a}_2$, and $\vec{a}_3$ are considered as (distinct) points of the projective plane, the line on any two is a line of fixed points. A collineation with two lines of fixed points must be the identity.*

## Exercise 13.5

*Solution: For parts (2)–(5), e-value 2 has e-vectors $(1, 2, 0)$ and $(0, 1, 1)$, so the axis is $[2, -1, 1]$, or $y = 2x + 1$. $(-1, 0, 1)$ is mapped to $(-1, -1, 0)$, so the vanishing line is on $(-1, 0)$ with slope 2. e-value 1 has e-vector $(-1, 1, 2)$, so $(-1/2, 1/2)$ is the center of the central collineation.*

*For part (6), the vanishing line meets the unit circle at $(-1, 0)$ and at $(-.6, .8)$; tangents to the unit circle at those points meet at $(-1, .5)$. This point is mapped to $(.5, .5)$. So $(.5, .5)$ is the center of the conic, a hyperbola.*

*Alternatively, we can find the pole of the vanishing line, which is mapped to the center of the conic.*

*Yet another way to find the center of the conic is to find two diameters of the conic. Lines $x = 0$ and $y = -2x + 2$ meet on the vanishing line. Line $y = -2x + 2$ meets the unit circle at $(1, 0)$ and $(.6,.8)$. These points are mapped to $(-1/8, 3/8)$ and $(-1/24, 15/24)$, whose midpoint is $(-1/12, 1/2)$, while $(0, 1)$ and $(0, -1)$ are mapped to $(0, 1)$ and $(-1/3, 0)$, whose midpoint is $(-1/6, 1/2)$. So a diameter of the conic is $y = 1/2$. Line $y = 2x + 1$, the axis, meets the unit circle at $(0, 1)$, which is fixed, and at $(-.8, -.6)$, also fixed. The midpoint is $(-.4, .2)$. Axis $y = 2x + 1$ is parallel to the vanishing line. Line $y = 2x - 2$ is on $(1, 0)$. It is parallel to the vanishing line and meets the unit circle at $(.6, -.8)$, which is mapped to $(-9/40, 7/40)$. As $(1, 0)$, on line $y = 2x - 2$, is mapped to $(-1/8, 3/8)$. Another diameter lies on $(-.4, .2)$ and the midpoint of segment from $(-9/40, 7/40)$ to $(-1/8, 3/8)$ : $(-7/40, 11/40)$. These points lie on $y = x/3 + 1/3$. Center: $(1/2, 1/2)$.*

## Exercise 13.6

*Solution of b.: $AC = [0, 1, 0]$, $DE = [1, 1, -2]$, and they meet at $(2, 0, 1)$; $CB = [1, 0, 0]$, $EF = [1, 1, -1]$, and they meet at $(0, 1, 1)$; $BD = [1, 0, -1]$, $AF = [0, 2, -1]$, and they meet at $(2, 1, 2)$. By the determinant, these three points are collinear. Why does this argument prove Pappus's Hexagon Theorem? Given three collinear points, we can name them by triples A, B, and $E = A - B$, and three other collinear points with none on line AB can be names C, D, and $F = C + D$. We know there is a non-singular 3-by-3 matrix mapping A, B, C, D to $(1, 0, 0)$, $(0, 1, 0)$, $(0, 0, 1)$, $(1, 1, 1)$, respectively, and those images will simply be called A, B, C, D. Matrix multiplication is linear so $E = A - B$ is mapped to $(1, -1, 0)$ and $F = C + D$ is mapped to $(1, 1, 2)$. Matrix multiplication is a collineation of the projective plane, so the argument above, that opposite sides of the hexagon meet in collinear points, holds for the original six points.*

# References

1. H. S. M. Coxeter, *Introduction to Geometry*, New York: Wiley, 1969.
2. A. Moebius, *Der Barycentrische Calcul: ein neues Hülfsmittel zur analytischen Behandlung der Geometrie*, Leipzig: Barth, 1827.
3. Daniel Pedoe, Notes on the history of geometrical ideas I. Homogeneous coordinates, *Mathematics Magazine*, Vol 48, No. 4, 1975, 215–217.
4. Julius Plücker, *Analytisch-Geometrische Entwicklungen*, two volumes, Essen: Baedeker, 1828, 1831.
5. K. G. C. von Staudt, *Geometrie der Lage*, Nürnberg: Bauer und Raspe, 1847; second edition, 1856–1857.

# Chapter 14
# Some Applications of Conics and Collineations in History

## 14.1 Archimedes' *Quadrature of the Parabola*

Archimedes (287–212 BC) is the most accomplished and best-known mathematician and scientist of the ancient world. He seems to have been the first to develop and apply mathematical formulations to physical phenomena, including the most basic principle of engineering: to balance weights $W_1$ and $W_2$ on a weightless lever when they are on opposite sides of the fulcrum, their distances from the fulcrum $d_1$ and $d_2$ must satisfy $W_1 d_1 = W_2 d_2$. His mathematical achievements include the discovery and proof of the volume formula for a sphere, and a method that can approximate $\pi$, the ratio of circumference to diameter of a circle, to any desired accuracy.

That he is best known—notice that we give his dates of birth and death with some confidence—is because facts of his life and career were recorded in works that survive, particularly in Plutarch's *Life* of the Roman general Marcellus, written around 75 AD. When Marcellus besieged Syracuse, Archimedes' city, on Sicily, from both land and sea, the Syracusans defended their city with the machines of Archimedes.

> huge poles thrust out from the walls over the ships sunk some by the great weights which they let down from on high upon them; others they lifted up into the air by an iron hand or beak like a crane's beak and, when they had drawn them up by the prow, and set them on end upon the poop, they plunged them to the bottom of the sea; or else the ships, drawn by engines within, and whirled about, were dashed against steep rocks that stood jutting out under the walls, with great destruction of the soldiers that were aboard them. [Plutarch, Life of Marcellus, [5]]

Plutarch reported that Archimedes was killed after the Romans entered the city, too involved in a geometric diagram to notice or obey the soldier who confronted him.

Archimedes found area and volume and center of mass for a number of figures, including those defined by conic sections. We give one example which depends on basic properties. Although Archimedes worked before Apollonius, the properties

© Springer Nature Switzerland AG 2020
C. Baltus, *Collineations and Conic Sections*,
https://doi.org/10.1007/978-3-030-46287-1_14

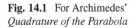

**Fig. 14.1** For Archimedes'
*Quadrature of the Parabola*

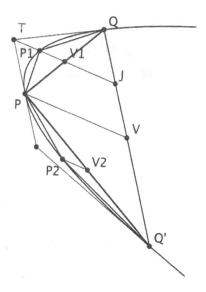

developed in the early books of the *Conics* were widely known in the time of
Archimedes. Our example, taken from *The Quadrature of the Parabola*, [2, pp. 248–
252], is one of the two derivations which show that the area of a segment of a
parabola is four-thirds the area of the largest triangle contained in that segment.

Let us recall that the diameters of a parabola are parallel, and that when a
tangent, at a point $Q$, to a parabola meets, at $T$, the diameter with vertex $P_1$, then
$P_1T = P_1V_1$, where $V_1Q$ is the corresponding ordinate on $Q$. [Theorem 6.3 *b*] See
Fig. 14.1. Further, the tangent at the vertex of a diameter is parallel to the ordinates
(Theorem 4.2). Further still, when $P$ and $Q$ are points on the parabola, and $V_1$ is the
midpoint of $PQ$ and the tangents at $P$ and $Q$ meet at $T$, then the line on $V_1$ and $T$
is another diameter (Theorem 6.2).

So when $PV$ is a diameter of the parabola, $PT$ the tangent at $P$, $QQ'$ an ordinate
corresponding to diameter $PV$, $V_1$ the midpoint of $PQ$, and $P_1V_1$ another diameter,
then since, further, $TJVP$ is a parallelogram, $TV_1$ is half of $PV$ and $P_1V_1$ is half
$TV_1$. Therefore the area of $\triangle PQP_1$ is a quarter that of $\triangle PQV$. In the same way,
the area of $\triangle PQ'P_2$ is a quarter that of $\triangle PQ'V$.

Now, let us step back to see the process Archimedes employed to find the area of
the parabolic segment with base $QQ'$. We build a sequence of inscribed polygons,
$PQQ'$, then $PP_1QQ'P_2$, etc., where from one to the next we add onto each side
other than $QQ'$ the largest triangle inscribed in the parabolic segment on that side.
At each step, the area added to the inscribed polygon is one-fourth of that added at
the previous step. So the $n^{th}$ polygon has the area of the first times $[1 + \frac{1}{4} + (\frac{1}{4})^2 +
\ldots + (\frac{1}{4})^n]$.

Now, we know the sum of the finite geometric series $[1 + \frac{1}{4} + (\frac{1}{4})^2 + \ldots + (\frac{1}{4})^n]$
is

$$\frac{1 - (\frac{1}{4})^{n+1}}{1 - \frac{1}{4}},$$

which converges to 4/3 as $n \to \infty$. Archimedes, by the practice of the times, dealt in area and not a sum of real numbers, and letting $n$ approach $\infty$ was not anything a Greek would do. He first showed by a sort of induction argument, that

$$1 + \frac{1}{4} + (\frac{1}{4})^2 + \ldots + (\frac{1}{4})^n = \frac{4}{3} - \frac{1}{3 \cdot 4^n}.$$

Then we first suppose the sum of areas was some value $K$ that is less than $4A/3$, where $A$ is the area of the first triangle; but this is impossible since $A[\frac{4}{3} - \frac{1}{3 \cdot 4^n}]$ eventually exceeds $K$. And if we suppose area $K$ is greater than $4A/3$, we see this is impossible since the inscribed polygons, always less in area than the parabolic segment, eventually differ in area from the area of the segment by less than $K - 4A/3$.

We conclude that the area of the parabolic segment is $4A/3$.

## 14.2  An Islamic Sundial

The Eastern Roman Empire, which included Alexandria, Egypt, and most of the Greek speaking world, survived the collapse of the Western Roman Empire, which occurred soon after 400 AD. But the Greek cultural and mathematical heritage was no longer supported, and much of it was lost over the centuries. The library at Alexandria, which had survived burning and pillaging several times, seems finally to have succumbed to a 391 AD edict to close all the pagan temples in Alexandria. (The library was the *museum*, dedicated to the muses, the goddesses of the arts.) Islam, which rose and spread rapidly in the seventh century AD, became in the ninth century a promoter of scholarship. A library, called the House of Wisdom, was established around 800 AD at the new city of Baghdad, and scholars and manuscripts were brought there from throughout the Middle East, as far east as India and central Asia. Among the early scholars at the House of Wisdom was al-Khwarizmi (c.780–850); he introduced Hindu numerals and arithmetic to the West, and wrote a book, whose terminology gives us the modern term "algebra," on solving first and second degree equations.

Among those works that barely survived was the *Conics* of Apollonius. It was copied and studied by Islamic scholars, who were eager to apply the mathematics of Greece to practical problems.

We'll look at one of these practical problems, the construction of sundials, through the writing of Ibrahim ibn Sinan (909–946). Sundials were common throughout the  Greek and Roman worlds, in which Islam spread in the seventh century. Muslims had a particular need to know the time of day: they were called to

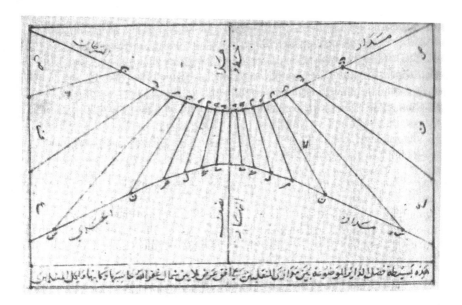

**Fig. 14.2** Sundial diagram, fifteenth century, latitude of Cairo, Egyptian National Library, in [3, p 88]

prayer at noon and in the afternoon halfway between noon and sunset. Thus, many mosques had sundials.

Ibn Sinan was from a family of doctors and scholars; the more famous Thabit ibn Qurra, his grandfather, had written an earlier work on sundials. What Ibn Sinan contributed was a more general treatment of the various types of sundials and, of most interest to us, a practical method of constructing points on a hyperbola to be etched into the metal plate of the sundial (Fig. 14.2).

In his autobiography, Ibn Sinan divides his works into those on astronomy and those on geometry. Among the first group was *Book on the Instruments for Shadows*, in which he gave the needed theory behind the construction of sundials. He wrote the first version as a teenager, then revised it at age 25. We have only part of this work. [6, p 291]

He began the first chapter of the first book with

> Our goal in this chapter is to describe and explain the question of curves traced by the edge of shadows on all planes. But since the known planes are the horizon, the meridian circle, the circle of the equator ... we need to know what happens on the horizon of all points located from the equator to the North Pole.... . [p 344]

Ibn Sinan first explained the path of the sun on the celestial sphere. The earth is considered to be a ball, so negligibly small that it could be thought of as nearly a point, at the center of a sphere—the celestial sphere on which the sun, planets, and stars moved. And every point $E$ on earth corresponds to a vector from the center of earth, $C$, which is also the center of the celestial sphere, and the horizon

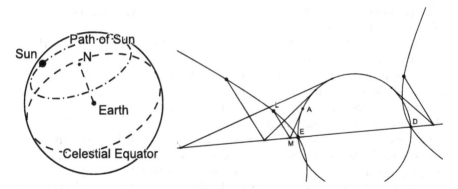

**Fig. 14.3** Left: The celestial sphere. The sun's height above the celestial equator is exaggerated. Right: Hyperbola construction (Ibn Sinan) with ML = MA

plane at each point of the earth is perpendicular to that vector. The vectors from the earth's equator together meet the celestial sphere in a great circle, called the celestial equator.

For us, now, we think of the earth rotating in the course of a day on its axis $\vec{CN}$, where $N$ is the North Pole, but it is the same to think of the earth as not rotating but, rather, the celestial sphere rotating about $\vec{CN}$. Let $S$ denote the sun on the celestial sphere. Through the daily rotation of the celestial sphere about $\vec{CN}$, since angle $SCN$ is constant, then the sun traces a circle, parallel to the celestial equator, on the celestial sphere . See Fig. 14.3 Left.

How does this look from a position $E$ on the earth? As Ibn Sinan wrote [6, p 344], "we treat the tip of the gnomon as the center of the celestial sphere." If we have a sundial with a flat plate and above the plate the tip of a gnomen (the piece on the plate, supporting the tip), where the shadow of that tip is to be followed on the plate, then the sun's rays in the course of a day form a conic surface for which the tip is the vertex and the sun's path on the celestial sphere is the circular base of the conic surface. The plate is a slicing plane to the conic surface. If the plate is in the horizontal plane at $E$, and $E$ is below the Arctic Circle, then the horizontal plane slices the cone in a hyperbola. There are also "vertical sundials," in which the cutting plane is perpendicular to the horizontal plane, giving, likewise, a hyperbola as the path of the shadow of the tip. (There are various models of sundials, in which the shape and placement of the gnomen is important. Here all that matters is the position of tip. Above the arctic circle, during part of the year, the conic section is an ellipse on the horizontal plane, or, at one particular time, a parabola. [p 368])

There is an exception at the equinoxes, when the sun's path is along the celestial equator. The sun's rays no longer form a cone, and the shadow follows a straight line on the plate of the sundial. [p 344, 346]

And, of course, different hyperbolic paths are produced on different days of the year, an issue addressed by Ibn Sinan, which we will not discuss. If we are to look at the points of the shadow at, say, 10 AM throughout the year, the path of those points is close to, but not precisely equal to, a straight line.

Now, how is a hyperbola to be traced on a flat plate? We turn to a work on geometry, *On the tracing of the three sections*, also written while Ibn Sinan was young and revised at age 25. The hyperbola is the third conic covered: "we find tracing it is possible by the circle and by lines." [p 274] We give the first of three methods. The hyperbola is specified by its diameter $DE$, by the angle, $\angle EML$, at which the ordinates meet the diameter, and by a positive constant.

First the circle with diameter $ED$ is drawn, where $E$ and $D$ are vertices of the conic on the diameter. Let $M$ be a point on the diameter extended past $E$. A tangent to the circle is drawn from $M$, meeting the circle at $A$. See Fig. 14.3 Right. (Construction of the tangent can be by Euclid Book 3 Prop. 17.) Then $L$ is marked on the ray from $M$, drawn at the given angle $EML$, so $ML = MA$. As $M$ ranges along the diameter extended, then the points $L$ give one side of a branch of the hyperbola. (Lengths $ML$ could all be multiplied by a constant.)

Why do we have a hyperbola? By Euclid's Book 3 Prop. 36, the Power-of-a-Point Theorem, our Theorem 3.3, $MA^2 = ME \cdot MD$. Now by Eq. (4.3), of Chap. 4, the hyperbola formed in the manner of Apollonius has equation

$$ML^2 = ME \cdot MD \cdot Constant1 = ME \cdot MD \cdot \frac{p}{ED}$$

where $p$ is the *parameter* or *latus rectum*. By the construction, $MA = ML$. Thus, Ibn Sinan's method gives us the hyperbola with the given diameter, the given angle of the ordinates to the diameter, and where the *latus rectum* has length $ED$.

## 14.3   Central Collineations in Perspective Drawing: Brook Taylor and G. J. 's Gravesande

Although representation of depth in painting goes back to Roman times, at least, systematic methods date only from the Renaissance. And mathematical theory trailed behind practice; according to Kirsti Andersen, the first "general theory of perspective" was by Guido del Monte, in 1600 [1, p 160].

Now, when a floor pattern, for example, appears in a painting, this is a projection from the painter's eye of the plane of the floor onto the plane of the canvas, or the plane of the floor is projected to the plane of the canvas projection. We will apply the second idea. The principles of a central collineation are behind several eighteenth century books.

Most notable of these were Brook Taylor's 1715 *Linear perspective or, a new method of representing justly all manner of objects as they appear to the eye, etc.* [7] and his related work [8], followed by more than a century of reworked editions. Brook Taylor is the mathematician who gave us "Taylor Series"; his work

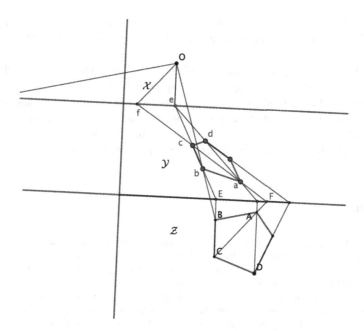

**Fig. 14.4** Based on Taylor's *Fig.* 16, 1715

on perspective was preceded by the *Essai de perspective*, a similar work of 1711 [4] by G. J. 's Gravesande. 's Gravesande was known for explaining Isaac Newton's mathematical physics to continental Europe.

Let us look at [7] *Fig.* 16, our Fig. 14.4 Left, and the figure to the right of it. Taylor's *Fig.* 16 is divided by parallel lines into three sections, labeled $X, Y, Z$. $Z$ represents the floor plane, on which we have points such as $B$ and $C$. $Y$ is the picture plane, the plane of the canvas, which meets the plane of the floor in the line where $Y$ and $Z$ meet. Plane $X$ is parallel to the plane $Z$ and contains the painter's eye, $O$. Line $ef$, the meeting of planes $X$ and $Y$, is the *vanishing line*, as Taylor called it, and $EF$ is the *axis*—not Taylor's name. Figure 14.4 Right displays the three planes in space.

Now we consider the construction of points $c$ and $b$, the preimages—points mapped to—$C$ and $B$, respectively. We have a *center*, $O$, which Taylor called the "spectator's eye." To locate $c$ by Taylor's basic method, we take two lines on $C$, $CB$ and $CA$, meeting the axis at $E$ and $F$, respectively. To find points, besides $E$ and $F$, which are mapped to lines $CE$ and $CF$, we find the preimages of the points at infinity on lines $CE$ and $CF$. They must lie on the vanishing line. We draw on $O$ parallels, $Oe$ and $Of$, to $CE$ and $CF$, respectively. $e$ is mapped to the point at infinity on line $CE$, so line $eE$ is mapped to line $CE$. In the same way, line $fF$ is mapped to line $FC$. Thus, $c$ is $eE \cap fF$.

The construction of $b$ follows another plan. As in locating $c$, $B$ is on $CE$ so $b$ is on $eE$. Since our mapping is a central collineation, $b$ lies on line $OB$, so $b$ is $OB \cap eE$.

Taylor did not apply rules of central collineations. His explanation as to why $b$ is on $OB$ is set out in Fig. 14.4 Right, following [8, p 27–30]. Planes $Oef$ and $BEF$ are parallel, and if we rotate those planes about fixed lines $ef$ and $EF$, keeping the planes parallel, then $b$ is unchanged: since lines $Oe$ and $BE$ stay parallel, then lines $Oe$, $eE$, and $BE$ are coplanar and, by similar triangles, point $b = OB \cap eE$ says unchanged. Until, as in Taylor's *Fig.* 16, the three planes become one.

# References

1. Kirsti Andersen, *The Geometry of an Art: The History of the Mathematical Theory of Perspective from Alberti to Monge*, New York: Springer, 2007.
2. Archimedes, *The Works of Archimedes with the Method of Archimedes*, ed T. L. Heath, Cambridge University press, 1912, reissued by New York: Dover, 2002.
3. J. L. Berggren, *Episodes in the Mathematics of Medieval Islam*, New York: Springer, 1986.
4. G. J. 'sGravesande, *Essai de perspective*, The Hague: veuve d'Abraham Troyel, 1711.
5. Plutarch, Plutarch's Lives, vol. 1, translation by John Dryden vol 1, New York: Cosimo, 2009.
6. Roshdi Rashad, Hélène Bellosta, editors, *Ibrahim ibn Sinan, Logique et géométrie au $X^e$ siècle*, Leiden, Netherlands: Brill, 2000.
7. Brook Taylor, *Linear Perspective: or a new method of representing justly all manner of objects as they appear to the eye in all situations*, London: Knaplock, 1715.
8. Brook Taylor, *New Principles of Linear Perspective*, London: Knaplock, 1719.

# Chapter 15
# Vertical Stretch and Isaac Newton's Conics

## 15.1 The Vertical Stretch

A *vertical stretch* is a central collineation whose axis is a horizontal line and whose center is at infinity on the lines perpendicular to the axis. When the axis is the *x-axis*, and point $(0, 1)$ is mapped to $(0, k)$, for $k \neq 0$, then each point $(x, y)$ is mapped to $(x, ky)$. The *scale factor* is $k$.

Properties: (1) The midpoint of a segment is mapped to the midpoint of the image segment. (2) If parallel segments have lengths in ratio $m/n$, then their images under a vertical stretch have lengths in the same ratio. (3) In mapping a bounded region with a "nice" boundary, the area is multiplied by $|k|$. (4) Parallel lies are mapped to parallel lines.

To see Property (2), note that the ratio of lengths of parallel segments is the ratio of the lengths of their projections onto the horizontal axis, and these projections are unchanged by a vertical stretch.

To see Property (3), note that any thin horizontal rectangle with length $w$ and height $\delta y$, is mapped to a thin horizontal rectangle of length $w$ and height $|k|\delta y$. The "nice" property means that the area is approximated as closely as desired by thin inscribed horizontal rectangles; since all those thin rectangles have their areas multiplied by $|k|$, the area of the region undergoes the same multiplication.

Now the ellipse with equation

$$\frac{x^2}{a^2} + \frac{y^2}{b^2} = 1 \tag{15.1}$$

is the image of circle $x^2 + y^2 = a^2$ under the vertical stretch with scale factor $b/a$.

Thus, conjugate axes of ellipse (15.1) are images of perpendicular diameters of the circle $x^2 + y^2 = a^2$, for all chords parallel to one of those diameters, which are bisected by the perpendicular diameter, are mapped to parallel chords bisected by the image of the perpendicular diameter. And conversely.

© Springer Nature Switzerland AG 2020
C. Baltus, *Collineations and Conic Sections*,
https://doi.org/10.1007/978-3-030-46287-1_15

## 15.2   Two Ellipse Properties

Isaac Newton's study of elliptical paths of planets about the sun made use of two important properties that he justified simply by referring to the *Conics* of Apollonius. We'll provide proofs in Theorems 15.1 and 15.2. The first, essentially the equation of the ellipse, follows from Book 1 Prop. 13 of Apollonius; we have demonstrated it in Chap. 4. The second is Prop. 31 of Book 7 of the *Conics*; Newton presented it without proof as Lemma 12 of Section 2, 1687.

Our proofs are based on the idea that an ellipse results from a vertical stretch applied to a circle.

**Theorem 15.1** *Let an ellipse with center C have diameter GP and conjugate diameter DK. Let Qv be an ordinate at v on diameter GP. Then*

$$\frac{Gv \cdot Pv}{Qv^2} = \frac{CP^2}{CD^2}.$$

***Proof*** See Fig. 15.1. The claim holds for conjugate diameters of a circle, i.e., perpendicular diameters, for by the power-of-point $v$, $Gv \cdot Pv = Qv^2$, and $CP^2 = CD^2$. Conjugate diameters of an ellipse are images of perpendicular diameters of a circle. $Gv$, $Pv$, $CP$ are parallels, as are $Qv$ and $CD$, so the ratios $Gv \cdot Pv/CP^2$ and $Qv^2/CD^2$ are unchanged under the vertical stretch that maps a circle to the given ellipse. So the same equation holds in the ellipse.                                          $\square$

**Fig. 15.1**   Theorem 15.1

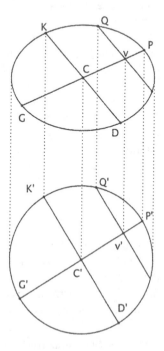

**Theorem 15.2** *In an ellipse with diameter $GP$ and conjugate diameter $DK$, the parallelogram formed by the tangents at $D, P, K, G$ has the same area no matter the choice of the conjugate diameters.*

*Proof* The claim is true for a circle, since conjugate diameters are, for a circle, perpendicular diameters, and the parallelograms involved are squares whose side is the diameter of the circle. Then by Property (3), above, under a vertical stretch that maps the perpendicular diameters of the circle to conjugate diameters of an ellipse, areas are multiplied by the scale factor, $|k|$, so the parallelograms described have equal areas.            □

## 15.3 Isaac Newton and the *Principia* of 1687

Isaac Newton (1642–1727) joins Archimedes among the most important figures in the history of both mathematics and physics. However, unlike Archimedes, his impact in physics was profound and immediate, and affected the worldview far beyond the small number of scientists who were able to follow his arguments. What had been suspected and hypothesized by that small number of scientists was clearly and indubitably demonstrated in Newton's *Mathematical Principles of Natural Philosophy*, of 1687: that the same law of gravity that governed free fall near the surface of the earth reigned also in the heavens, with the pull of the sun at one focus holding the bodies of the solar system in their elliptical orbits.

Born in Lincolnshire, England, Newton was raised by a grandmother when his widowed mother remarried and moved to live with her new husband. On the death of her second husband, the mother returned to the family farm and two years later, in 1655, Isaac was sent off to grammar school at Grantham, where he learned Latin and some arithmetic. His years at school were interrupted by a return to manage the farm, which he thoroughly disliked. In 1662 he entered Trinity College of Cambridge University, as a subsizer, a poor student who covered expenses by services to wealthier students.

The Cambridge curriculum at the time was antiquated and little enforced. Newton learned from his reading, although he probably attended the year of mathematics lectures of the Lucasian Professor, Isaac Barrow. We have several notebooks produced by Newton in his university years, in which he made notes of his wide-ranging reading and his own reflections. He was interested in light and vision, the motion of bodies, and the nature of the universe. He taught himself the mathematics of the seventeenth century, especially the coordinate geometry of Descartes and John Wallis's exploration of patterns. He wrote in 1699 [4, p 98] that he had developed the method of infinite series by 1665, "And in summer 1665 being forced from Cambridge by the Plague I computed ye area of ye Hyperbola at Boothby in Lincolnshire to two and fifty figures by the same method." (He was referring to the series for the natural logarithm.)

**Fig. 15.2** Newton
manuscript on circular
motion, c. 1669

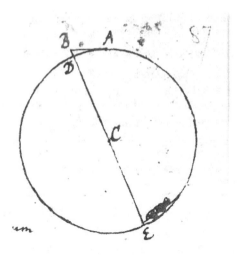

Newton was made a "scholar" in 1664, which provided a stipend for at least
four years, allowing him to continue at Cambridge when it reopened in 1667. The
Newton who returned in 1667 was not the same scholar. On the farm for the two
years of the plague, he developed rules for what we now recognize as calculus, he
studied conic sections, and he mastered algebra beyond anything found in books of
the time. He worked out the physics of force, impact, and momentum, if not yet with
the clarity of his 1687 work. Richard Westfall writes that Newton developed a "new
definition of force in which a body was treated as the passive subject of external
forces impressed upon it instead of the active vehicle of force impinging on others"
[4, p 146].

Johannes Kepler had announced, based on precise observation of the planets,
his first two Laws in 1609: that planets move in elliptic orbits for which the sun
is a focus, and the line from sun to planet sweeps out equal areas in equal times.
The Third Law, published in 1618, states that the ratio $r^3/T^2$ is a constant for the
planets, where $r$ is the radius of the orbit and $T$ the period, or duration, of the orbit.

We follow Stuart Hollingdale's simplifying summary [1, p 195] of Newton's
early work on uniform circular motion. The diagram of Fig. 15.2 is from a
manuscript of about 1669 [2, p 297]. We assume constant speed $|v|$ for velocity
$v$. In time $t$ the body moves distance $vt$ in a straight line, $AB$. And under constant
acceleration $a$, a body starting at rest has velocity $at$, and, therefore, falls distance
$BD = s = at^2/2$ in time $t$. When the body moves along the circle of radius $r$ from
$A$ to $D$ it is pulled toward the center, $C$, of the circle through distance $s$. Now, by the
power-of-point for point $B$, $BA^2 = s(s + 2r)$. $s$ is negligible compared to $r$, so we
get the approximate equation $(vt)^2 = 2rs$. Replacing $s$ by $at^2/2$ gives $a = v^2/r$.
When $T$ is the time required for one complete revolution, then $vT = 2\pi r$, so
$$a = \frac{4\pi^2}{r^2}\frac{r^3}{T^2}.$$

By Kepler's Law $r^3/T^2$ is constant, from which it follows that the force that holds a body in circular orbit of radius $r$ is proportional to $1/r^2$. As Newton wrote at the time, [4, p 152] "the endeavours of receding from the Sun [he discovered] will be reciprocally as the squares of the distances from the Sun."

Using the $v^2/r$ formula, Newton found that the force of gravity on the surface of the earth was about 4,000 times that needed to hold the moon in a circular orbit about the earth, agreeing approximately with the inverse square law for the force of gravity. He would later prove that the earth's gravity could be considered to operate from the center point of the earth.

Soon after returning to Trinity College, Cambridge, in 1667, Newton became a fellow of the college. A work of his on infinite series, *De analysi*, was made known in scientific circles due to the efforts of Isaac Barrow and John Collins, who was in communication with prominent mathematicians. In 1669, Newton succeeded Isaac Barrow as Lucasian Professor, a post that gave him security and a comfortable living.

Over the next fifteen years, Newton devoted himself more to biblical studies, optics and light, and alchemy/chemistry than to the physics of motion or mathematics, although he did some work on algebra and curves that he refused to publish. In April 1684, he was visited in Cambridge by Edmond Halley. As reported [1, p 174] by Abraham de Moivre in 1727, transcribed by John Conduitt, "after they [Halley and Newton] had been some time together, the Dr. asked him what he thought the Curve would be that would be described by the Planets supposing the force of attraction towards the Sun to be reciprocal to the square of their distance from it. Sir Isaac replied immediately that it would be an Ellipsis, the Doctor struck with joy and amazement asked him how he knew it. Why saith he I have calculated it." Unable to find his old calculation, Newton promised to renew it and send it to Halley.

Newton set to work with his remarkable intensity. By the end of 1684 he delivered to Halley a short study *De motu corporum—On the movement of bodies*— that proved to be a preliminary version of the first of the three books of the *Principia*. The full book was finished in late 1686, published the next year.

What did Newton prove about elliptic orbits? In just Proposition 1 of Section 2, Kepler's Second Law was proved. It did not require an elliptical orbit or an inverse square law; if the orbit of a body is determined by a central force, then the line from the point source to the body sweeps out equal areas in equal time. The crucial part of the argument is in the following theorem.

**Theorem 15.3 (Equal Area)** *Suppose a body moves unaffected by external force from A to B in time interval t, and then while at B is subjected to a "single but great impulse" toward S, resulting in its movement from B to C over an equal time interval t. Then triangles ABS and BCS have equal areas. See Fig. 15.3, from Newton 1687.*

***Proof*** Let $BV$ indicate the movement of the body due exclusively to the impulse while the body is at $B$, and $Bc$ is the movement due to the momentum (not Newton's term) in direction $AB$, "if nothing hindered it." So $BC$ is the vector sum of those

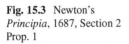

**Fig. 15.3** Newton's
*Principia*, 1687, Section 2
Prop. 1

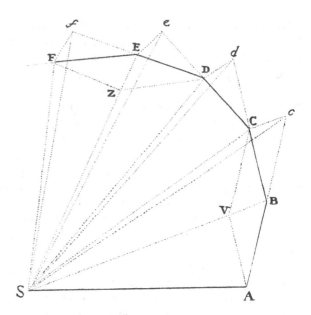

two movements; $AB$ and $Bc$ are equal vectors, and $VB$ and $Cc$ are equal vectors. Since $B$ is the midpoint of base $Ac$ of triangle $AcS$, triangles $ABS$ and $BcS$ have equal areas. And since triangles $BcS$ and $BCS$ have the same base, $SB$, and equal heights, they have equal areas. Therefore, triangles $ABS$ and $BCS$ have equal areas.

$\square$

To prove Kepler's second law, Newton simply noted that the smooth path of a body subjected only to a central force was the limit of polygonal paths $ABCDEF$ over shorter and shorter time intervals $t$.

Then in a sequence of propositions, he proved that if a body moved on an elliptical path under the influence of a central force at a focus of the ellipse, then the force obeys the inverse square law.

We first give Newton's Section 1 Lemma 11 and a corollary [3, p 439]. See Fig. 15.4 Left.

**Theorem 15.4** *Suppose, as pictured, curve $AbB$, at point $A$, has center of curvature $J$, with $DB$ perpendicular to tangent $AD$. Then as $B$ approaches $A$ along the curve, ratios $BD/AB^2$ and $BD/(arcAB)^2$ approach a finite non-zero limit. (Newton: BD "is ultimately in the squared ratio of the subtense [AB] of the conterminous arc.") Further, if $t$ is the time in which the body traces $arcBA$ and the ratio (arclength AB)/t has a non-zero finite limit as $BA \to 0$, then $BD/(arcAB)^2$ has a finite non-zero limit.*

*Modern Proof* Since angle $BAD$ intercepts arc $AB$, then $sin(\angle BAD) = sin(\dfrac{arcAB}{2}) = BD/AB$. Divide through by $(arcAB)/2$, then take the limit as $AB \to 0$. This gives

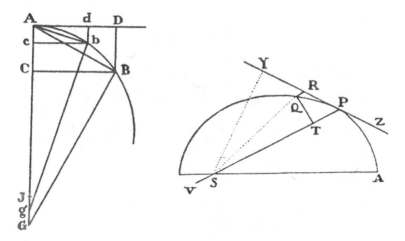

**Fig. 15.4** Left: Newton's *Principia* Section 1 Lemma 11. Right: Newton's *Principia* Section 2 Prop. 6

$$1 = lim \frac{2BD}{AB \cdot arcAB} = lim \frac{2BD}{(arcAB)^2}.$$

Then Prop. 6 and a corollary, of Section 2, relate the central force causing deflection from a straight line path to the elapsed time and to associated lengths.

**Theorem 15.5 (Section 2 Prop. 6 Corollary, 1687)** *Consider the path $APQ$ of a body under a central force $F$ at focus $S$, traveling from $P$ to $Q$ in time $t$. See Fig. 15.4 Right. With $R$ on the tangent from $P$, where $QR \parallel SP$ and $QT \perp SP$, then the force $F$ is proportional with* $\dfrac{QR}{SP^2 \cdot QT^2}$.

**Proof** Galileo had shown that a body starting at rest travels distance $at^2/2$ in time $t$ under acceleration $a$. So the force $F$, proportional to the acceleration, is proportional to $QR/t^2$. By the Equal Area Theorem, the area of $\triangle PQS$, $\frac{1}{2}SP \cdot QT$, is proportional to $t$. The theorem immediately follows. □

Newton's major claim in this part of the *Principia* is Prop. 11 of Section 2. In addition to the theorems above in this chapter, we make use of two well-known conic properties that involve the focus. These are parts $a$. and $b$. of our Theorem 9.3, found in Book 3 of the *Conics* of Apollonius:

$a$. Lines from the foci to a point of tangency meet the tangent in equal angles.
$b$. The sum of the lengths of these line segments is always the length of the axis.

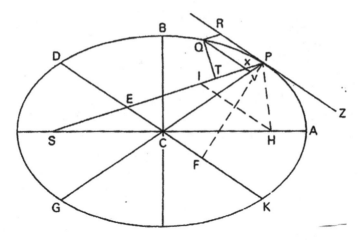

**Fig. 15.5** Left: Newton's *Principia* Section 2 Prop. 11

Here is Prop. 11, quoted:

**Theorem 15.6 (Newton's *Principia*, 1687, Section 2 Prop. 11)** *Let a body revolve in an ellipse; it is required to find the law of centripetal force tending toward a focus of the ellipse [3, p 462].*

*Newton's First Proof* See Fig. 15.5. $S$ is a focus of the ellipse with center $C$, where $H$ is the other focus, and point $P$ is on the ellipse. $PG$ and $DK$ are conjugate diameters. We draw $RPZ$ tangent at $P$, and join $S$ and $P$. $Q$ is another point on the ellipse, near to $P$; $QR$ is parallel to $SP$ and $QT$ is the perpendicular to $SP$. $HI$ is drawn parallel to $PR$, and $PF \perp DK$. $Qxv$ is drawn parallel to $PR$, meeting $SP$ at $x$ and $CP$ at $v$. $SP \cap DK = E$.

i. By Property $a$. above, $\angle RPS \cong \angle ZPH$, from which it follows that $IP = HP$.
ii. Since $C$ is the midpoint of $SH$ and $EC \parallel IH$, then $SE = EI$.
iii. By Property $b$. above, $SE + EI + IP + PH = 2CA$. By $i$. and $ii$., $EP = CA$.
iiii. By parallels, $Px/Pv = PE/PC$, so $QR/Pv = AC/PC$.
v. By Theorem 15.1, $\dfrac{Gv \cdot Pv}{Qv^2} = \dfrac{CP^2}{CD^2}$.
vi. As $Q$ approaches $P$, then since $PR$ is tangent, $Qx/Qv \to 1$, so we can replace $Qv$ by $Qx$.
vii. By parallels, $\angle QxE \cong \angle PEF$ so $\triangle QxT \sim \triangle PEF$, so $\dfrac{QT}{Qx} = \dfrac{PF}{PE}$.

Then

$$\frac{QR}{QT^2} = \frac{AC \cdot Pv}{PC} \frac{1}{QT^2} = \frac{AC \cdot Pv}{PC} \frac{PE^2}{PF^2 \cdot Qx^2} = \frac{AC \cdot Pv}{PC} \frac{PE^2}{PF^2 \cdot Qv^2} =$$

$$\frac{AC \cdot Pv}{PC} \frac{PE^2}{PF^2} \frac{PC^2}{CD^2 \cdot Gv \cdot Pv} = \frac{AC \cdot PE^2}{BC^2 \cdot AC^2} \frac{CP}{Gv}$$

where the last equality is by Theorem 15.2. Since $PE = AC$, then $\dfrac{AC \cdot PE^2}{BC^2 \cdot AC^2}$ is a constant. Conclude: as $Q$ nears $P$ along the ellipse, $QR/QT^2$ is proportional to $CP/Gv$. Then Newton wrote, "with the points $Q$ and $P$ coming together, 2 $PC$ and $Gv$ are equal" [3, p 463]. So $QR$ and $QT^2$ are proportional.

Finally, we know from Theorem 15.5 that the force from $S$ is proportional to $\dfrac{QR}{SP^2 \cdot QT^2}$. Therefore the force is "inversely as the square of the distance $SP$."

Newton then proved the converse. In the 1687 edition he simply noted the converse, but by the 1726 edition, he gave complete proofs, in Book 1, Prop 13., with corollaries.

## 15.4   A Conic Construction of Isaac Newton

Sections 4 and 5 of Book 1 of the *Principia* are devoted to constructing conics with given information. This work was done before Newton's work on the *Principia* [3, p 136]. Section 4 is introduced, "To find elliptical, parabolic, and hyperbolic orbits, given a focus." His results extend what is found in Apollonius. Section 5 is introduced "To find orbits when neither focus is given."

We present just one of Newton's constructions, in Book 5 Lemma 21 and Prop. 22. It is readily justified by the Steiner definition of a conic as the set of points of intersection of two projectively related pencils. See Fig. 15.6.

We are given five points of a conic, $A$, $B$, $C$, $D$ and $P$. We need two fixed angle measures: $m\angle ABC$ at $B$, and $m\angle ACB$ at $C$. We draw $DB$ and $DC$. Then we draw rays $BM$ and $CM$, meeting at $M$, so $\angle DBM = \angle ABC$ and $\angle DCM = \angle ACB$. Likewise, draw rays $BN$ and $CN$ so $\angle PBN = \angle ABC$ and $\angle PCN = \angle ACB$. Draw line $MN$. Then we find additional points on the conic on $A, B, C, D$ and $P$ by following the relationship of $B, C, N$ and $P$: for any point $N$ on line $MN$, draw $NB$ and $NC$. Then at $B$ draw angle $NBP$ congruent to angle $ABC$ and at $C$ draw angle $NCP$ congruent to angle $ACB$. Rays $CP$ and $BP$ meet at a point, $P$, of the conic.

Why do we have a conic? The pencils on $B$ and $C$ are paired projectively with corresponding lines meeting on line $MN$. Then the pencil on $B$ is rotated by angle $ABC$ and the pencil on $C$ is rotated by angle $ABC$, producing pencils which are still projectively related. Among the points at which corresponding lines meet are the five given points $A, B, C, D, P$.

Note that we have, by this construction, a method to produce, point by point, the conic on five given points.

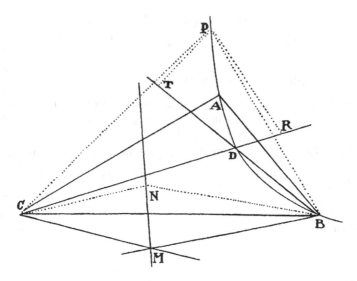

**Fig. 15.6**  Newton Section 5 Prop. 22, 1687

## 15.5  Notes

The diagrams from Newton's *Principia* are from the 1726 edition, as given in [3].
The same text gives a thorough introduction to this work of Newton. For a thorough,
yet accessible, presentation of Newton's career and the most important points of
his *Principia*, I recommend [1, p 167–227]. Finally, there is Richard Westfall's
monumental scientific biography, *Never at Rest* [4].

## References

1. Stuart Hollingdale, *Makers of Mathematics*, Harmondsworth, England: Pelican Books, 1994.
2. Isaac Newton, *The Correspondence of Isaac Newton*, vol 1, edited by H. W. Turnbull, Cambridge
   University Press, 1959.
3. Isaac Newton, *The Principia: Mathematical Principles of Natural Philosophy*, edited and
   translated by I. B. Cohen and Anne Whitman, based on the 1726 third edition, in Latin. Berkeley:
   University of California Press, 1999.
4. Richard Westfall, *Never at Rest, A Biography of Isaac Newton*, Cambridge University Press,
   1980.

# Index

© Springer Nature Switzerland AG 2020
C. Baltus, *Collineations and Conic Sections*,
https://doi.org/10.1007/978-3-030-46287-1

Printed in the United States
by Baker & Taylor Publisher Services